De Gruyter Graduate

Kazmierczak, Azzazy • Diagnostic Enzymology

Also of Interest

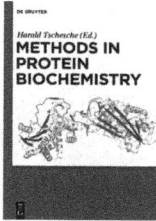

Methods in Protein Biochemistry
Harald Tschesche (Ed.), 2011
ISBN 978-3-11- 025233-0, e-ISBN 978-3-11- 025236-1,

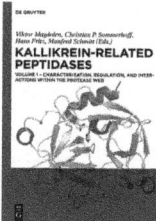

Kallikrein-Related Peptidases – Volume 1 Characterization, regulation, and interactions within the protease web
Viktor Magdolen, Christian P. Sommerhoff, Hans Fritz, Manfred Schmitt, 2012
ISBN 978-3-11- 026036-6, e-ISBN 978-3-11- 026037-3,

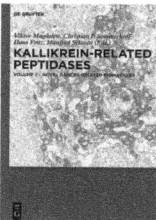

Kallikrein-Related Peptidases – Volume 2 Novel cancer-related
Viktor Magdolen, Christian P. Sommerhoff, Hans Fritz, Manfred Schmitt, 2012
ISBN 978-3-11- 030358-2, e-ISBN 978-3-11- 030366-7,

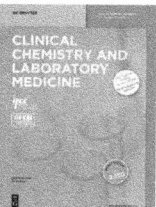

Clinical Chemistry and Laboratory Medicine
Mario Plebani (Editor-in-Chief)
ISSN 1437-4331

Statistical Applications in Genetics and Molecular Biology
Michael P.H. Stumpf (Editor-in-Chief)
ISSN 1544-6115

www.degruyter.com

Diagnostic Enzymology

Edited by
Steven C. Kazmierczak, Hassan M. E. Azzazy

2nd fully revised and extended edition

DE GRUYTER

Editors

Prof. Dr. Steven C. Kazmierczak
Oregon Health & Science University
Department of Pathology
3181 SW Sam Jackson Park Road
Portland, OR 97239
USA
kazmierc@ohsu.edu

Prof. Dr. Hassan M.E. Azzazy
The American University in Cairo
School of Science and Engineering
Department of Chemistry
AUC Avenue, P.O. Box 74
11835 New Cairo
Egypt
hazzazy@aucegypt.edu

ISBN 978-3-11-020724-8
e-ISBN 978-3-11-022780-2

Library of Congress Cataloging-in-Publication data
A CIP catalog record for this book has been applied for at the Library of Congress.

Bibliographic information published by the Deutsche Nationalbibliothek
The Deutsche Nationalbibliothek lists this publication in the Deutsche Nationalbibliografie; detailed
bibliographic data are available in the Internet at http://dnb.dnb.de.

© 2014 Walter de Gruyter GmbH, Berlin/Boston

Typesetting: Compuscript Ltd, Shannon, Ireland
Printing and binding: CPI buch bücher.de GmbH, Birkach
Cover image: Ovidiu Iordachi, Getty Images/Hemera
∞ Printed on acid-free paper
Printed in Germany

www.degruyter.com

Preface

The field of clinical diagnostic enzymology has undergone some significant changes since the first edition of the textbook, *Clinical Enzymology: A Case Oriented Approach*, was published by Lott and Wolf in 1986. Better standardization of methods used to measure enzymes and a greater understanding of *in vivo* and *in vitro* factors that affect the interpretation of changes in measured enzyme values necessitated an update to the original version of the text. A new chapter on natriuretic peptides, which have significant clinical utility for diagnosis of heart failure, has been included.

Our goal for this textbook was to update and expand upon the original text in order to provide a world class resource for students in clinical pathology as well as to help educators teach students the nuances of interpretation of changes in enzyme values.

Each chapter begins with a series of case studies that highlight some of the complexities of diagnostic enzymology. Following the case studies in each chapter are detailed discussions of the biochemistry and physiology of each enzyme and the role that each enzyme plays in health and disease. Reference intervals for each enzyme are also supplied. However, due to the fact that reference values for many enzymes are method dependent, the reference values that are supplied may not be applicable to all methods and analytical instruments.

Each chapter ends with a number of multiple choice questions which readers can use to assess their knowledge of clinical enzymology. These questions should also help readers preparing for board-type examinations. The question and solution manual can be accessed online at http://dx.doi.org/10.1515/9783110227802_Exercise.

We would like to thank the many authors who provided the excellent source material for the chapters. We would also like to thank Ms. Diana Abou El-Hassan and Ms. Heba Othman for their editorial assistance. Finally, we are in debt to our families for their forbearance and support throughout this project.

April 2014

Steven C. Kazmierczak, Ph.D., DABCC
Hassan M.E. Azzazy, Ph.D., DABCC

List of contributing authors

Hassan M.E. Azzazy
Department of Chemistry
The American University in Cairo
School of Science and Engineering
New Cairo, Egypt
email: hazzazy@aucegypt.edu
Chapter 10

Sarah Brown
Department of Pathology and
Immunology
Division of Laboratory and Genomic Medicine
Washington University
St. Louis, MO, USA
e-mail: brown_sa@kids.wustl.edu
Chapter 7

Robert Christenson
University of Maryland Medical Center
Baltimore, MD, USA
e-mail: rchristenson@umm.edu
Chapter 10

Dennis Dietzen
Washington University School of
Medicine
St. Louis, MO, USA
e-mail: dietzen_d@kids.wustl.edu
Chapter 8

Joe M. El-Khoury
Department of Clinical Pathology
Cleveland Clinic
Cleveland, OH, USA
e-mail: joe.eldouss@gmail.com
Chapter 1, 3, 5, 9

Olajumoke Oladipo
Staten Island University Hospital
Staten Island, NY 10305
e-mail: ooladipo@siuh.edu
Chapter 8

Peter L. Platteborze
Medical Service Corps, US Army
Department of Laboratory Medicine
University of Washington
Seattle, WA, USA
e-mail: Peter.platteborze@us.army.mil
Chapter 2

Edmunds Reineks
Department of Clinical Pathology
Cleveland Clinic
Cleveland, OH, USA
e-mail: reineke@ccf.org
Chapter 1, 3, 9

Amy E. Schmidt
Department of Pathology and Laboratory Medicine
Indiana University
Indianapolis, IN, USA
e-mail: amyschm@iupui.edu
Chapter 4

Sihe Wang
Department of Clinical Pathology
Cleveland Clinic
Cleveland, OH, USA
e-mail: wangs2@ccf.org
Chapter 1, 3, 5, 9

Alan Wu
Departments of Pathology and Laboratory
Medicine
University of California at San Francisco
San Francisco, CA, USA
e-mail: wualan@labmed2.ucsf.edu
Chapter 6

Wan-Ming Zhang
Department of Clinical Pathology
Cleveland Clinic
Cleveland, OH, USA
e-mail: zhangw2@ccf.org
Chapter 9

Contents

8   Lactate dehydrogenase —— 135
Olajumoke Oladipo and Dennis J. Dietzen

9   Pancreatic lipase —— 153
Wan-Ming Zhang, Edmunds Reineks, Joe M. El-Khoury and Sihe Wang

1 Angiotensin converting enzyme

Joe M. El-Khoury, Edmunds Reineks and Sihe Wang

1.1 Case studies

1.1.1 Patient A

A 55-year-old White male presented with fatigue, dizziness, pressure headaches, and facial numbness. Initial evaluations revealed some neurological abnormalities, including bilateral hand pain, progressive dizziness, and increased sensitivity to sound. His medications included meclizine, a headache medicine containing butalbate, and topiramate.

Neurological evaluation 3 months after onset of symptoms revealed blunted facial sensation and decreased vibratory sensation in his finger tips and toes. Position sensation was intact. A magnetic resonance imaging (MRI) of the brain showed mild, non-specific white matter changes most likely representing microvascular disease. However, demyelinating disease could not be excluded. Carotid ultrasound studies were normal.

Analysis of cerebrospinal fluid (CSF) showed a normal angiotensin converting enzyme (ACE) <14 U/L, and increased protein of 0.69 g/L (elevated), and negative for oligoclonal bands.

1.1.1.1 Discussion

Neurosarcoidosis is a rare manifestation that occurs in 5% of patients with systemic sarcoidosis. Lesions mainly involve the optic and cranial nerves and the spinal cord. This patient presented with fairly non-specific neurological complaints. The differential diagnosis includes various inflammatory, infectious, and autoimmune disorders, including neurosarcoidosis and multiple sclerosis. The low ACE level and lack of oligoclonal bands in the CSF made these diagnoses less likely, and the patient was scheduled for additional follow-up. One month after the evaluation the patient reported considerable improvement and discontinued his various medications. Based on the laboratory findings, most of the pathological processes could be excluded and given the patient's improvement, further follow-up was not deemed necessary.

1.1.2 Patient B

A 41-year-old African-American female who was known to have sarcoidosis with pulmonary involvement presented for evaluation of worsening dyspnea, blurred vision, and recurring headaches. Her diagnosis of sarcoidosis was made 2 years previously by mediastinoscopy, which revealed the presence of non-caseating granulomas. At this visit, her pulmonary function tests were normal and there was no evidence of hypoxemia. Her only medication was hydroxychloroquine, which is useful for the treatment of sarcoidosis, particularly that of the skin.

Laboratory evaluation showed a normal cell blood count (CBC), except for elevated eosinophils (7%), and a normal metabolic panel. Her serum ACE was mildly elevated (53 U/L, normal <49 U/L) and her soluble interleukin-2 receptor (sIL-2R) levels were normal. The patient was prescribed a course of prednisone and was scheduled for a follow-up appointment.

1.1.2.1 Discussion

This patient's disease activity was evaluated and, on the basis of her test results, was not severe. The mildly elevated ACE level was consistent with the established diagnosis of sarcoidosis. Measurement of sIL-2R levels has been suggested to play a role as a marker for pulmonary disease activity, but there is no relationship between sIL-2R levels and response to treatment. This patient's sarcoidosis seemed to be properly controlled by hydroxychloroquine, but of concern were the various side effects caused by this drug, which include headaches, blurred vision, and more serious ocular toxic effects. Given this patient's increased symptomology and her recent emergency department visits, her hydroxychloroquine therapy was transiently stopped and her condition improved with prednisone therapy.

1.1.3 Patient C

A 41-year-old female was referred to the clinic by her primary care physician for uveitis, parotitis, and mediastinal lymphadenopathy. The patient had been previously healthy but had a worsening cough over the past 3 months, along with swollen knees, ankles, and parotids with dry mouth. In addition, the patient had unintentionally lost significant weight (18 kg) and had recurring fevers 39°C with night sweats. She was given a short course of antibiotics by her primary care physician but developed diffuse erythematous rash 1 week after stopping the antibiotics.

Physical examination was normal. Her laboratory results were unremarkable except for elevated calcium (3.3 mmol/L), aspartate aminotransferase (49 U/L), alkaline phosphatase (348 U/L), and ACE (195 U/L).

1.1.3.1 Discussion

This patient presented with a multisystemic process involving the mediastinal lymph nodes, uveitis, cough, fever, and weight loss in the past 3 months. The possible diagnoses included malignancy (non-Hodgkin's lymphoma), sarcoidosis, or fungal infections. On the basis of the elevated ACE and calcium levels, sarcoidosis seemed to be the most likely diagnosis. In fact, the symptoms described are most consistent with

Heerfordt's syndrome, a rare manifestation of sarcoidosis, which involves uveitis, parotiditis, and chronic fever. This diagnosis was further confirmed by bronchoscopy findings. Steroid treatment was initiated and the patient's condition improved, but her ACE level remained elevated (>100 U/L) during follow-up visits.

1.2 Biochemistry and physiology

1.2.1 Physiological function

Angiotensin I converting enzyme (ACE; EC 3.4.15.1) is a type-I membrane-anchored zinc metallopeptidase of the M2 family with diverse physiological functions [1]. Isolated for the first time in 1956, ACE is known to be mainly involved in blood pressure regulation and electrolyte hemostasis through the renin-angiotensin system (RAS) by acting as a dicarboxypeptidase in bradykinin inactivation (Figure 1.1a) and angiotensin II production (Figure 1.1b) [2]. Bradykinin is a vasodilator whereas angiotensin II is a vasoconstrictor. As a result, ACE promotes vasoconstriction and is a target for treating hypertension and associated cardiovascular disorders [3]. Other substrates of this enzyme include angiotensin (1–7), hemoregulatory peptide N-Acetyl-Ser-Asp-Lys-Pro, gonadotropin-releasing hormone, luteinizing hormone-releasing hormone, substance P, enkephalins, β-neoendorphin (1–9), and neurotensin [4–6].

ACE2 or hACE, a human homolog of ACE and a new component of the RAS system, was discovered in 2000 by two independent groups [7, 8]. It has been demonstrated that ACE2 also acts as a carboxypeptidase but only cleaves a single amino acid from the C-terminus of angiotensin I to yield angiotensin (1–9) and angiotensin II to yield angiotensin (1–7) (Figure 1.1c) [7]. Among these two pathways of the so-called "alternative RAS," cleavage of angiotensin II is 400-fold more kinetically favored over that of angiotensin I [9–11]. Animal studies have shown that angiotensin (1–7) has anti-hypertensive, anti-arrhythmic, anti-inflammatory, antifibrotic, antithrombotic, anti-trophic, and cardioprotective properties [11]. Therefore, the protective effects of the "alternative RAS" axis counter-regulate the deleterious effects of the "classical RAS" axis, thereby preventing overactive RAS-associated diseases, including hypertension [12]. Currently, ACE2 is still under research and is not used for diagnostic purposes. As a result, the remainder of this chapter will focus on ACE only.

1.2.2 Biochemistry and molecular forms

ACE is expressed as two isoforms, somatic (sACE) and testis (tACE), derived from a common gene by alternative splicing. The larger somatic isoform (1306 AA, 150 kDa) consists of N- and C-terminal catalytic extracellular domains, each containing the zinc-binding motif HEMGH (His-Glu-Met-Gly-His), whereas tACE (732 AA, 83 kDa) consists of a single extracellular domain almost identical with the C domain of

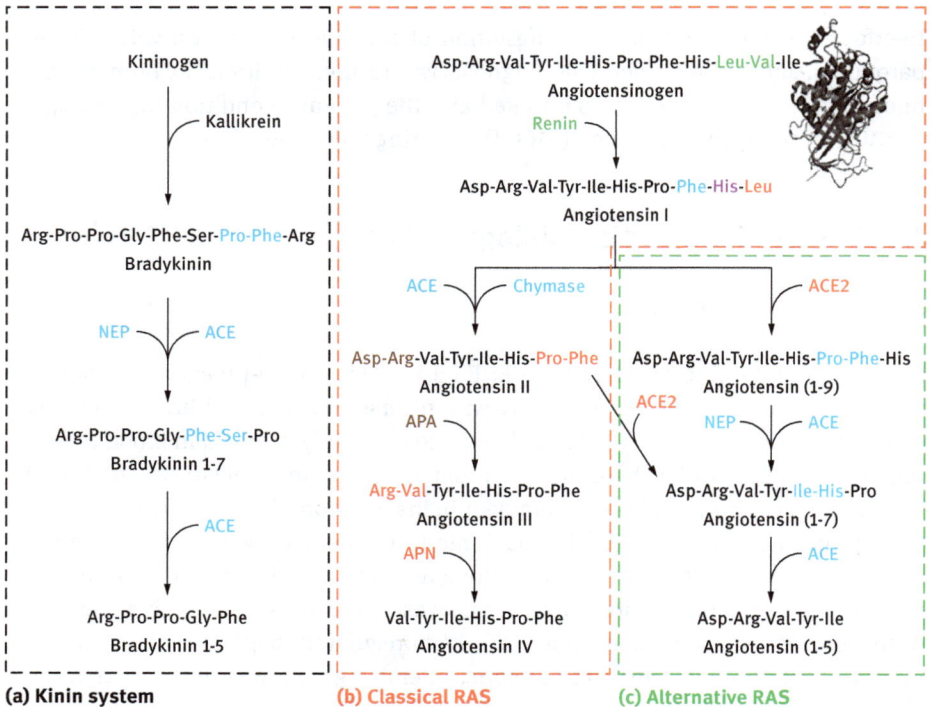

Figure 1.1: Schematic illustration highlighting the enzymes, their substrates, and their sites of cleavage in the kinin system (a), classical renin-angiotensin system (RAS) (b), and alternative RAS (c). Colored amino acids represent the site of cleavage for the color-matched enzymes. ACE, angiotensin converting enzyme; ACE2, angiotensin converting enzyme 2; NEP, neprilysin; APA, aminopeptidase A; APN, aminopeptidase N.

sACE [13, 14]. Differences in structure and function of the N and C domains of sACE imply that the two active sites each have independent activity.

Although ACE has been known for some time, obtaining a crystal structure of this enzyme has proved challenging due to variation in surface glycosylation, rendering it difficult to crystallize. As a result, the first X-ray crystal structure of human tACE complexed with lisinopril, a commonly used ACE inhibitor, was not reported until 2003 after modification of the carbohydrates to minimize oligosaccharide-based heterogeneity [15]. The structure of tACE was reported to be an overall ellipsoid shape with a central groove dividing the molecule into two subdomains. Covering the top of the molecule is an N-terminal "lid" formed by three α-helices that is believed to restrict access of large polypeptide to the active site cleft. This structure is predominantly helical with 27 helices and only one β structure located near the active site in six relatively short strands that account for 4% of all residues.

Both domains of ACE are heavily glycosylated, with the C domain containing seven and the N domain containing ten potential N-glycosylation sites [16]. Investigations into the role of glycosylation in the N domain have revealed that C-terminal glycosylation is vital for folding and processing, whereas N-terminal glycosylation maintains its high level of thermal stability [17]. It is suggested that this higher degree of glycosylation, combined with a greater number of α-helices and increased proline content, contributes to its thermal stability (T_m = 70°C) compared with the C domain (T_m = 55°C) [16, 18]. Functionally, although the two active sites of sACE are able to cleave angiotensin I and bradykinin *in vitro*, the C domain is the major site of angiotensin I cleavage *in vivo* [19–23]. In addition, the two domains exhibit differences in chloride activation and specificity for various inhibitors [20, 24, 25].

Chloride-dependent enzyme activation is an unusual characteristic of ACE that is shared only by a handful of other enzymes [26]. The C domain of sACE requires chloride in a pH-dependent manner for its catalytic action on angiotensin I, whereas the N domain retains 45% of its catalytic activity on bradykinin in the absence of chloride [27]. In addition, maximal substrate turnover is observed at 200 mM of chloride for the C domain, but at only 20 mM of chloride for the N domain, which also decreases at chloride concentrations greater than 20 mM [28]. The crystal structure of tACE reveals the location of two buried chloride ions separated by 20.3 Å [15]. Activation of ACE by other monovalent anions is also possible, with chloride and bromide having the highest affinity, whereas iodide and fluoride have a ten- to 100-fold lower affinity [29]. The order of activation of ACE by monovalent anions is chloride > bromide > fluoride > nitrate > acetate. The mechanism of ACE chloride activation is hypothesized to be through stabilization of the enzyme-substrate complex by inhibition of salt bridge formation between R522 and D465 because ion pairing between R522 and the chloride ion allows Y523 to move towards the catalytic center and stabilize the complex [30].

1.2.3 Tissue sources

The lung has been recognized as a rich source of ACE, but the enzyme has also been isolated from over 20 other tissues in humans [31]. ACE activity in blood is mostly of endothelial origin [32]. In the kidney and intestine, ACE is present in both endothelial and epithelial cells [33] with the highest ACE activity present in the kidney, ileum, and duodenum, which is more than two-fold of that found in the lung [31]. Seven tissues, including jejunum, prostate, lymph node, thyroid, colon, testis, and adrenal, display similar ACE activity as the lung, whereas 12 other tissues have lower activity than the lung, with the heart consistently showing the least ACE activity. Cells derived from the monocyte/macrophage system also contain ACE [33].

1.2.4 Reference ranges

In most clinical investigations, serum ACE activity has been largely found to be independent of age (in adults greater than 18 years), gender, or race [33]. Nevertheless, reference range determinations for serum ACE activity remain a huge challenge because of poor assay standardization and genetic polymorphism [34].

As described in more detail in Section 4, there are a variety of assays used for the measurement of serum ACE activity, with the most routinely used today being the automated kinetic assays that employ N-[3-(2-furyl)acryloyl)]-L-phenylalanylglyclglycine (FAPGG) as the substrate. However, reference intervals determined using different methods with FAPGG as the substrate are fairly dissimilar [35]. This discrepancy was hypothesized to be caused by instrument-specific differences in the measurement of the absorbance change and a means of correction was proposed [36]. However, there are still several other factors affecting assay performance, including wavelength selection, calibration, substrate molarity, buffers, sample volumes, and lag phase. These factors led to unacceptable between-laboratory variation in the interpretation of ACE results in a study performed in the UK [34]. As a result, it is recommended that every laboratory establish its own reference intervals. A summary of reference intervals for various methods employing FAPGG as the substrate are summarized in Table 1.1.

Genetic polymorphism is another factor confounding reference range establishment. The insertion/deletion (*I/D*) of a 287 base pair fragment in intron 16 of the ACE gene on chromosome 17, in which either allele can be *I* or *D*, gives rise to three possible ACE genotypes: *I/I*, *D/D*, and *D/I* [37]. D/D individuals have nearly twice the serum ACE activity of I/I individuals, with I/D individuals being somewhere in between [34]. In a study involving 159 healthy Caucasian volunteers, genotype-corrected reference values were determined for serum ACE using two different kinetic tests, one using FAPGG (assay 1) and the other His-His-Leu (assay 2) as the substrate [38]. The upper limit of the normal range without correcting for genotype was 81 U/L and 62 U/L using assays 1 and 2, respectively. After correcting for genotype, the reference values were 30–89

Table 1.1: Reference intervals for spectrophotometric assays employing FAPGG as the substrate.

Reference	Reference interval (U/L)	n
Buttery and Chamberlain [36]	32–105	54
Biller et al. [38]	All, 12–82	159
	D/D, 30–89	47
	D/I, 16–75	73
	I/I, 8–62	39
Ronca-Testoni [39]	35–155	66
Maguire and Price [40]	20–95	100
Buttery and Stuart [41]	41–139	133
Neels et al. [42]	33–147	103
Harjanne [43]	35–155	66
Johansen et al. [44]	14–110	92
Beneteau et al. [45]	30–170	150
Hendriks et al. [46]	19–93	97
Roulston and Allan [47]	11–115	100

and 16–71 for *D/D*, 16–75 and 10–57 for *D/I*, and 8–62 and 7–44 for *I/I* using assays 1 and 2, respectively. Although using genotype-corrected reference values may enhance the diagnostic accuracy, the requirement to simultaneously perform polymerase chain reaction (PCR) analysis for each patient limits its widespread use as a screening tool.

1.3 Chemical pathology

Abnormal ACE activities have been reported in a variety of diseases, as summarized in Table 1.2. However, laboratory testing of ACE plays a limited role in the diagnosis and patient management of sarcoidosis or monitoring the effectiveness of ACE inhibitor therapy in hypertensive patients.

1.3.1 ACE in sarcoidosis

Sarcoidosis is a systemic inflammatory disorder of unknown etiology that commonly involves epithelioid granulomas primarily affecting the lungs, although all parts of

Table 1.2: Conditions with increased or decreased serum ACE activity.

Increased serum ACE	Decreased serum ACE
Sarcoidosis [33]	Fibrosis (idiopathic, cryptogenic alveolitis, and cystic) [33]
Active tuberculosis [33]	Acute respiratory distress syndrome [33]
Atypical mycobacterium [33]	Chronic obstructive pulmonary diseases [33]
Miliary tuberculosis [33]	Bronchial carcinoma [33]
Leprosis [33]	Extrinsic allergic alveolitis [33]
Gaucher's disease [33]	Pulmonary tuberculosis [33]
Primitive biliary cirrhosis [33]	Wegener's granulomatosis [33]
Hodgkin's disease [33]	Crohn's disease [33]
Lennert's lymphoma [33]	Ulcerative colitis [33]
Lymphangioleiomyomatosis [33]	Extrahepatic obstruction [33]
Silicosis [33]	Hypothyroidism [33]
Asbestosis [33]	Acute renal failure [33]
Berylliosis [33]	Deep vein thrombosis [33]
Histoplasmosis [33]	Veno-occlusive disease [33]
Coccidioidomycosis [33]	Corticotherapy [33]
Hyperthyroidism [33]	Chemotherapy of leukemia [33]
Acute alcohol hepatic disease [33]	ACE inhibitors [33]
Diabetes mellitus [33]	Smoking [34]
Human immunodeficiency virus infection [34]	Hormone replacement therapy [34]
Chronic fatigue syndrome [34]	
Cardiovascular disease risk [34]	
Multiple sclerosis [34]	

the body may be affected. It is more prevalent in individuals 20–40 years of age, and its diagnosis relies on compatible clinical/radiographic presentation, histological evidence of non-caseating granulomas, and exclusion of known causes of granulomatous disease [48]. Its pharmacological treatment involves corticosteroid therapy.

The link between sarcoidosis and serum ACE was reported in 1975 by Lieberman, who also suggested that serum ACE may be useful for the diagnosis and management of this disease [49]. Elevation of serum ACE in sarcoidosis is believed to be caused by stimulation of the monocyte-macrophage system, with the epithelioid cells originating from macrophages being the most likely source of ACE [50]. As a diagnostic test, serum ACE activity is increased in 40% to 60% of patients with sarcoidosis, with positive and negative predicative values between 75% and 90% and 70% and 80%, respectively [49, 51, 52]. The specificity of this test for sarcoidosis is not high because many other granulomatous diseases (Table 1.2) may also cause elevation in serum ACE. Elevations greater than two times the upper limit of the reference range are much less common with diseases other than sarcoidosis, but unfortunately may also be seen in tuberculosis, Gaucher's disease, and hyperthyroidism [53]. In a study that examined the influence of ACE genotype on the interpretation of the diagnostic test for serum ACE activity, the diagnostic sensitivity improved from 47% to 57% to 65% to 70%, whereas the diagnostic specificity fell from 77% to 58% [54]. Furthermore, serum ACE activity does not correlate with chest radiographic stage at diagnosis [55], a test which detects abnormalities in 85% to 95% of patients with sarcoidosis [56]. However, ACE activity significantly correlates with the extent of nodules and consolidation on high resolution computed tomography [57], and may be useful when combined with other diagnostic tests. For example, when combined with [67]GA scintigraphy, the sensitivity of serum ACE is increased to almost 100% in radiographic type II or III patients, providing a high negative predictive value for excluding active pulmonary sarcoidosis [58]. In addition, serum ACE activity test is a useful prognostic tool for monitoring the course of sarcoidosis when a diagnosis has been firmly established [33]. A high initial serum ACE activity level portends a poor prognosis, whereas a normal or low level is indicative of a good prognosis. ACE activity tends to rise as the disease progresses, stays elevated in chronic active widespread disease or falls back towards normal as the disease remits spontaneously, or with corticosteroid treatment [33]. Nevertheless, owing to its limited sensitivity and specificity, the British Thoracic Society guidelines suggest that the serum ACE activity test has a limited role in diagnosis and does not contribute to staging or monitoring of patients with pulmonary sarcoidosis when added to serial lung function and imaging tests [59, 60].

In addition to serum, ACE activity measurement can be performed in CSF to aid in the diagnosis of neurosarcoidosis. However, this test has also shown limited sensitivity (55%), and the ACE level cannot be used to discriminate between neurosarcoidosis and other neuronal disorders [59, 61].

Table 1.3: Commercially available ACE inhibitors.

Drug generic name	Brand name
Benazapril	Lotensin
Captopril	Capoten
Cilazapril	Inhibace
Enalapril	Vasotec
Fosinopril	Monopril
Lisinopril	Prinivil, Zestril
Moexipril	Univasc
Perindopril	Aceon
Quinapril	Accupril
Ramipril	Altace
Trandolapril	Mavik

1.3.1.1 ACE inhibitors in hypertensive disease

ACE inhibitors have been widely used for the treatment of hypertension and patients at high risk for coronary artery disease, after myocardial infarction, with dilated cardiomyopathy, or with chronic kidney disease [62]. Ever since the development of the first ACE inhibitor, captopril, in the 1970s, other ACE inhibitors with more affective pharmacodynamic effects have become available (Table 1.3). Routine measurement of serum ACE activity in patients treated with ACE inhibitors may be advantageous in assessing medication compliance because most of the current ACE inhibitors completely inhibit serum ACE activity [33]. However, this practice has not been adopted in the medical community.

1.4 Analysis

1.4.1 Specimen

Serum or heparinized plasma can be used for the analysis of ACE activity. EDTA plasma cannot be used because EDTA chelates zinc and therefore has an inhibitory effect on this zinc-containing enzyme [49]. ACE activity in serum has been reported to be stable for at least 20 days at 25°C, 1 month at 4°C, and 6 months at −20°C [50]. Storage at −70°C or freezing and thawing have been reported to increase serum ACE activity by 15% [63].

A plethora of methods for determining ACE activity in serum and tissue have been reported, including bioassay, radioisotopic, fluorometric, capillary electrophoresis, enzyme immunoassay, high performance liquid chromatography (HPLC), and spectrophotometric methods, with the latter being the most widely adopted method in clinical laboratories because they are easily automated.

1.4.2 Spectrophotometric methods

Several spectrophotometric methods have been described for the measurement of ACE activity, among which the majority are kinetic assays that measure the disappearance rate of a substrate. The fundamental difference among these methods has been the substrates used, which have mostly been tripeptides. Cushman and Cheung described a method that involves ethyl acetate extraction and is based on the rate of hippurate release from hippuryl-L-histidyl-L-leucine monitored at 228 nm [64]. Several variations of this method have been reported in the literature [49, 65–68], of which some are interfered by hemoglobin and/or bilirubin [69]. Other substrates used include p-hydroxyhippuryl-L-histidyl-L-leucine [70, 71] and hippuryl-glycyl-glycine [72, 73]. FAPGG was found to be the most suitable among several furanacryloyl blocked tripeptides tested, and therefore is most used in modern assays [74]. ACE catalyzes the hydrolysis of FAPGG to furylacrylolylphenylalanine (FAP) and glycylglycine resulting in a blue shift of the absorption spectrum which is monitored at 340 or 345 nm. A method employing FAPGG as the substrate was shown to correlate closely (r = 0.995) with Cushman's modified method, whereas the rapidity, simplicity, sensitivity, and precision of the FAPGG method make it much more suitable for routine use [38]. In addition, use of FAPGG as the substrate provides a more reliable measure of change in ACE activity in patients receiving ACE inhibitors, as reflected by the angiotensin II/angiotensin I ratio [75]. As a result, many methods have been developed using FAPGG as the substrate [36, 40–42], and some have been adopted in numerous automated assays [43–46, 76].

1.4.3 High-performance liquid chromatography

Based on the spectrophotometric assay by Cushman and Cheung [64], a HPLC method to measure hippuric acid formation and consequently ACE activity was first introduced by Chiknas [77]. Modifications of this method have also been reported [78–83]. Similarly, other HPLC methods have been developed for the measurement of hippuric acid released from hippuryl-glycl-glycine [84] and p-hydroxyhippuric acid from p-hydroxyhippuryl-L-histidyl-L-leucine [70] that correlate well with the corresponding colorimetric assay (r = 0.92 and r = 0.986, respectively). In addition, a HPLC method using FAPGG as the substrate without extraction has been reported [85]. Notably, the HPLC methods circumvent several interference problems encountered in the spectrophotometric assays and offer improved sensitivity and specificity [86].

1.4.4 Fluorometric methods

Friedland and Silverstein reported two fluorometric assays for the measurement of ACE activity [87, 88]. The first uses hippuryl-L-histidine-L-leucine as the substrate [87], whereas the other measures the conversion of its natural substrate angiotensin I [88]. Both methods rely on the formation of a fluorescent adduct with o-phthalalde-hyde, which was measured spectrophotometrically. In another reported method, a weak fluorescent substrate o-aminobenzoylglycl-p-nitro-L-phenylalanyl-L-proline is enzymatically hydrolyzed to produce the highly fluorescent o-aminobenzoylglycine [89]. This substrate is also used in a direct kinetic fluorometric assay, which, however, was found unsuitable for icteric or hemolyzed samples due to interference [90]. Using a different approach, Kapiloff et al. [91] reported the use of fluorescently labeled tri-peptides (dansyl-phenylalanyl-arginyl-tryptophan or dansyl-phenylalanyl-arginyl-phenylalanine) as the substrates. Hydrolysis of the peptides by ACE results in the release of a dipeptide and dansyl-phenylalanine, which partitions quantitatively into chloroform allowing rapid and sensitive measurement of ACE activity [91].

1.4.5 Radioassays

Several methods that use radiolabeled substrates have been reported. Enzyme activity is measured via monitoring the release of radiolabeled hippuric acid from either p-[3H]benzoylglycylglycylglycine [92] or (glycine-1-[14C])hippurate-L-histidyl-L-leucine [93–95]. Methods using the latter as the substrate were found to be more sensitive and free from interference by hemolysis and lipemia [93].

1.4.6 Other methods

Other methods in the literature for determining ACE activity include an enzymatic assay [96] and a capillary zone electrophoresis-based assay [97]. In the enzymatic assay, ACE catalyzes a reaction that releases glycylglycine from hippuryl-glycyl-glycine. The glycylglycine then participates in an enzymatic reaction, catalyzed by γ-glutamyltransferase (GGT; EC 2.3.2.2), as the receptor substrate of the γ-glutamyl group from L-γ-glutamyl-3-carboxy-4-nitroanilide, forming 3-carboxy-4-nitroaniline, which is measured spectrophotometrically at 410 nm. The capillary zone electro-phoresis based assay separates the products of the ACE catalyzed hydrolysis of hippuryl-L-histidyl-L-leucine and the ACE activity was determined by quantification of hippuric acid monitored at 228 nm.

1.4.7 Inhibitors of ACE

EDTA and other compounds, which form complexes with zinc and inhibit the formation of the active metalloprotein, inhibit ACE activity. Cobalt, manganese, cadmium, mercury, lead, copper, and nickel significantly inhibit ACE activity at 0.2 mmol/L, whereas zinc enhances ACE activity by 96% at the same concentration [47]. Zinc is also able to restore activity lost due to EDTA chelation. Some synthetic peptides, sulfhydryl containing compounds, and chelating agents also significantly inhibit ACE activity as described by Cushman and Cheung [65].

1.5 Questions and answers

1. Through which pathway does ACE play a role in blood pressure regulation?
    (a) Angiotensin I production
    (b) Angiotensin II production
    (c) Bradykinin inactivation
    (d) Angiotensin I production and bradykinin inactivation
    (e) Angiotensin II production and bradykinin inactivation
2. Which statement is false about the two isoforms of ACE (sACE and tACE)?
    (a) They are derived from a common gene by alternative splicing
    (b) The somatic isoform (sACE) is larger than tACE
    (c) tACE is almost identical with the N domain of sACE
    (d) The crystal structure of tACE was revealed first in 2003
    (e) Both isoforms are heavily glycosylated
3. What does ACE require to maintain maximum activity?
    (a) Chloride
    (b) Zinc
    (c) Vitamin B6
    (d) Zinc and chloride
    (e) Zinc and vitamin B6
4. What factor limits the utility of ACE reference ranges?
    (a) Serum ACE activity is largely independent of age, gender, or race
    (b) Genetic polymorphism
    (c) Chloride-dependent activation
    (d) Discovery of ACE2
    (e) Poor clinical sensitivity
5. What is the most likely source of elevated serum ACE activity in sarcoidosis?
    (a) Epithelioid cells originating from macrophages
    (b) Lungs
    (c) Kidneys
    (d) Hepatic cells
    (e) Intestines

6. What may cause elevated levels of ACE activity?
   (a) Sarcoidosis
   (b) Neurosarcoidosis
   (c) Tuberculosis
   (d) Gaucher's disease
   (e) All of the above
7. Which specimen is unacceptable for the measurement of ACE activity?
   (a) Cerebrospinal fluid
   (b) Serum separator tube
   (c) Heparinized plasma
   (d) EDTA plasma
   (e) Plain serum tube
8. What are the most widely adopted methods for the measurement of ACE in clinical laboratories?
   (a) Spectrophotometric methods
   (b) High performance liquid chromatography methods
   (c) Fluorometric methods
   (d) Radioassays
   (e) Capillary electrophoresis

References

[1] Corradi HR, Schwager SL, Nchinda AT, Sturrock ED, Acharya KR. Crystal structure of the N domain of human somatic angiotensin I-converting enzyme provides a structural basis for domain-specific inhibitor design. J Mol Biol 2006,357,964–74.

[2] Skeggs LT Jr, Kahn JR, Shumway NP. The preparation and function of the hypertensin-converting enzyme. J Exp Med 1956,103,295–9.

[3] Heran BS, Wong MM, Heran IK, Wright JM. Blood pressure lowering efficacy of angiotensin converting enzyme (ACE) inhibitors for primary hypertension. Cochrane Database Syst Rev 2008,CD003823.

[4] Deddish PA, Marcic B, Jackman HL, Wang HZ, Skidgel RA, Erdos EG. N-domain-specific substrate and C-domain inhibitors of angiotensin-converting enzyme: angiotensin-(1-7) and keto-ACE. Hypertension 1998,31,912–7.

[5] Rousseau A, Michaud A, Chauvet MT, Lenfant M, Corvol P. The hemoregulatory peptide N-acetyl-Ser-Asp-Lys-Pro is a natural and specific substrate of the N-terminal active site of human angiotensin-converting enzyme. J Biol Chem 1995,270,3656–61.

[6] Sturrock ED, Natesh R, van Rooyen JM, Acharya KR. Structure of angiotensin I-converting enzyme. Cell Mol Life Sci 2004,61,2677–86.

[7] Donoghue M, Hsieh F, Baronas E, Godbout K, Gosselin M, Stagliano N, et al. A novel angiotensin-converting enzyme-related carboxypeptidase (ACE2) converts angiotensin I to angiotensin 1-9. Circ Res 2000,87,E1–9.

[8] Tipnis SR, Hooper NM, Hyde R, Karran E, Christie G, Turner AJ. A human homolog of angiotensin-converting enzyme. Cloning and functional expression as a captopril-insensitive carboxypeptidase. J Biol Chem 2000,275,33238–43.

[9] Vickers C, Hales P, Kaushik V, Dick L, Gavin J, Tang J, et al. Hydrolysis of biological peptides by human angiotensin-converting enzyme-related carboxypeptidase. J Biol Chem 2002,277,14838–43.

[10] Rice GI, Thomas DA, Grant PJ, Turner AJ, Hooper NM. Evaluation of angiotensin-converting enzyme (ACE), its homologue ACE2 and neprilysin in angiotensin peptide metabolism. Biochem J 2004,383,45–51.

[11] Lubel JS, Herath CB, Burrell LM, Angus PW. Liver disease and the renin-angiotensin system: recent discoveries and clinical implications. J Gastroenterol Hepatol 2008,23,1327–38.

[12] Xia H, Lazartigues E. Angiotensin-converting enzyme 2: central regulator for cardiovascular function. Curr Hypertens Rep 2010,12,170–5.

[13] Soubrier F, Alhenc-Gelas F, Hubert C, Allegrini J, John M, Tregear G, et al. Two putative active centers in human angiotensin I-converting enzyme revealed by molecular cloning. Proc Natl Acad Sci U S A 1988,85,9386–90.

[14] Ehlers MR, Fox EA, Strydom DJ, Riordan JF. Molecular cloning of human testicular angiotensin-converting enzyme: the testis isozyme is identical to the C-terminal half of endothelial angiotensin-converting enzyme. Proc Natl Acad Sci U S A 1989,86,7741–5.

[15] Natesh R, Schwager SL, Sturrock ED, Acharya KR. Crystal structure of the human angiotensin-converting enzyme-lisinopril complex. Nature 2003,421,551–4.

[16] O'Neill HG, Redelinghuys P, Schwager SL, Sturrock ED. The role of glycosylation and domain interactions in the thermal stability of human angiotensin-converting enzyme. Biol Chem 2008,389,1153–61.

[17] Anthony CS, Corradi HR, Schwager SL, Redelinghuys P, Georgiadis D, Dive V, et al. The N domain of human angiotensin-I-converting enzyme: the role of N-glycosylation and the crystal structure in complex with an N domain-specific phosphinic inhibitor, RXP407. J Biol Chem 2010,285,35685–93.

[18] Voronov S, Zueva N, Orlov V, Arutyunyan A, Kost O. Temperature-induced selective death of the C-domain within angiotensin-converting enzyme molecule. FEBS Lett 2002,522,77–82.

[19] Junot C, Gonzales MF, Ezan E, Cotton J, Vazeux G, Michaud A, et al. RXP 407, a selective inhibitor of the N-domain of angiotensin I-converting enzyme, blocks in vivo the degradation of hemoregulatory peptide acetyl-Ser-Asp-Lys-Pro with no effect on angiotensin I hydrolysis. J Pharmacol Exp Ther 2001,297,606–11.

[20] Watermeyer JM, Kroger WL, O'Neill HG, Sewell BT, Sturrock ED. Characterization of domain-selective inhibitor binding in angiotensin-converting enzyme using a novel derivative of lisinopril. Biochem J 2010, 428,67–74.

[21] Fuchs S, Xiao HD, Cole JM, Adams JW, Frenzel K, Michaud A, et al. Role of the N-terminal catalytic domain of angiotensin-converting enzyme investigated by targeted inactivation in mice. J Biol Chem 2004,279,15946–53.

[22] Fuchs S, Xiao HD, Hubert C, Michaud A, Campbell DJ, Adams JW, et al. Angiotensin-converting enzyme C-terminal catalytic domain is the main site of angiotensin I cleavage in vivo. Hypertension 2008,51,267–74.

[23] van Esch JH, Tom B, Dive V, Batenburg WW, Georgiadis D, Yiotakis A, et al. Selective angiotensin-converting enzyme C-domain inhibition is sufficient to prevent angiotensin I-induced vasoconstriction. Hypertension 2005,45,120–5.

[24] Wei L, Alhenc-Gelas F, Corvol P, Clauser E. The two homologous domains of human angiotensin I-converting enzyme are both catalytically active. J Biol Chem 1991,266,9002–8.

[25] Wei L, Clauser E, Alhenc-Gelas F, Corvol P. The two homologous domains of human angiotensin I-converting enzyme interact differently with competitive inhibitors. J Biol Chem 1992,267,13398–405.

[26] Pokhrel R, McConnell IL, Brudvig GW. Chloride regulation of enzyme turnover: application to the role of chloride in photosystem II. Biochemistry 2011,50,2725–34.

[27] Jaspard E, Wei L, Alhenc-Gelas F. Differences in the properties and enzymatic specificities of the two active sites of angiotensin I-converting enzyme (kininase II). Studies with bradykinin and other natural peptides. J Biol Chem 1993,268,9496–503.

[28] Buenning P, Riordan JF. Activation of angiotensin converting enzyme by monovalent anions. Biochemistry 1983,22,110–6.

[29] Liu X, Fernandez M, Wouters MA, Heyberger S, Husain A. Arg(1098) is critical for the chloride dependence of human angiotensin I-converting enzyme C-domain catalytic activity. J Biol Chem 2001,276,33518–25.

[30] Tzakos AG, Galanis AS, Spyroulias GA, Cordopatis P, Manessi-Zoupa E, Gerothanassis IP. Structure-function discrimination of the N- and C- catalytic domains of human angiotensin-converting enzyme: implications for Cl- activation and peptide hydrolysis mechanisms. Protein Eng 2003,16,993–1003.

[31] Lieberman J, Sastre A. Angiotensin-converting enzyme activity in postmortem human tissues. Lab Invest 1983,48,711–7.

[32] Drouet L, Baudin B, Baumann FC, Caen JP. Serum angiotensin-converting enzyme: an endothelial cell marker. Application to thromboembolic pathology. J Lab Clin Med 1988,112,450–7.

[33] Beneteau-Burnat B, Baudin B. Angiotensin-converting enzyme: clinical applications and laboratory investigations on serum and other biological fluids. Crit Rev Clin Lab Sci 1991,28,337–56.

[34] Muller BR. Analysis of serum angiotensin-converting enzyme. Ann Clin Biochem 2002,39, 436–43.

[35] Buttery JE. Why are there different reference intervals for the kinetic angiotensin-converting enzyme assay? Clin Chem 1987,33,1491–2.

[36] Buttery JE, Chamberlain BR. A scheme for determining the correct activity of the kinetic angiotensin-converting enzyme. Clin Chem 1985,31,645–6.

[37] Rigat B, Hubert C, Alhenc-Gelas F, Cambien F, Corvol P, Soubrier F. An insertion/deletion polymorphism in the angiotensin I-converting enzyme gene accounting for half the variance of serum enzyme levels. J Clin Invest 1990,86,1343–6.

[38] Biller H, Zissel G, Ruprecht B, Nauck M, Busse Grawitz A, Muller-Quernheim J. Genotype-corrected reference values for serum angiotensin-converting enzyme. Eur Respir J 2006,28,1085–90.

[39] Ronca-Testoni S. Direct spectrophotometric assay for angiotensin-converting enzyme in serum. Clin Chem 1983,29,1093–6.

[40] Maguire GA, Price CP. A continuous monitoring spectrophotometric method for the measurement of angiotensin-converting enzyme in human serum. Ann Clin Biochem 1985,22,204–10.

[41] Buttery JE, Stuart S. Assessment and optimization of kinetic methods for angiotensin-converting enzyme in plasma. Clin Chem 1993,39,312–6.

[42] Neels HM, Scharpe SL, van Sande ME, Fonteyne GA. Single-reagent microcentrifugal assay for angiotensin converting enzyme in serum. Clin Chem 1984,30,163–4.

[43] Harjanne A. Automated kinetic determination of angiotensin-converting enzyme in serum. Clin Chem 1984,30,901–2.

[44] Johansen KB, Marstein S, Aas P. Automated method for the determination of angiotensin-converting enzyme in serum. Scand J Clin Lab Invest 1987,47,411–4.

[45] Beneteau B, Baudin B, Morgant G, Giboudeau J, Baumann FC. Automated kinetic assay of angiotensin-converting enzyme in serum. Clin Chem 1986,32,884–6.

[46] Hendriks D, Scharpe S, van Sande M, Vingerhoed JP. Single-reagent automated determination of angiotensin converting enzyme in serum. Clin Chem 1985,31,1761.

[47] Roulston JE, Allan D. Studies with an automated kinetic assay for plasma angiotensin-converting enzyme activity and its potentiation by zinc ion. Clin Chim Acta 1987,168,187–98.

[48] Spagnolo P, Luppi F, Roversi P, Cerri S, Fabbri LM, Richeldi L. Sarcoidosis: challenging diagnostic aspects of an old disease. Am J Med 2012,125,118–25.

[49] Lieberman J. Elevation of serum angiotensin-converting-enzyme (ACE) level in sarcoidosis. Am J Med 1975,59,365–72.

[50] Studdy PR, Lapworth R, Bird R. Angiotensin-converting enzyme and its clinical significance – a review. J Clin Pathol 1983,36,938–47.

[51] Lieberman J, Nosal A, Schlessner A, Sastre-Foken A. Serum angiotensin-converting enzyme for diagnosis and therapeutic evaluation of sarcoidosis. Am Rev Respir Dis 1979,120,329–35.

[52] Studdy PR, Bird R. Serum angiotensin converting enzyme in sarcoidosis – its value in present clinical practice. Ann Clin Biochem 1989,26,13–8.

[53] Statement on sarcoidosis. Joint Statement of the American Thoracic Society (ATS), the European Respiratory Society (ERS) and the World Association of Sarcoidosis and Other Granulomatous Disorders (WASOG) adopted by the ATS Board of Directors and by the ERS Executive Committee, February 1999. Am J Respir Crit Care Med 1999,160,736–55.

[54] Stokes GS, Monaghan JC, Schrader AP, Glenn CL, Ryan M, Morris BJ. Influence of angiotensin converting enzyme (ACE) genotype on interpretation of diagnostic tests for serum ACE activity. Aust N Z J Med 1999,29,315–8.

[55] Shorr AF, Torrington KG, Parker JM. Serum angiotensin converting enzyme does not correlate with radiographic stage at initial diagnosis of sarcoidosis. Respir Med 1997,91,399–401.

[56] Keir G, Wells AU. Assessing pulmonary disease and response to therapy: which test? Semin Respir Crit Care Med 2010,31,409–18.

[57] Leung AN, Brauner MW, Caillat-Vigneron N, Valeyre D, Grenier P. Sarcoidosis activity: correlation of HRCT findings with those of ^{67}Ga scanning, bronchoalveolar lavage, and serum angiotensin-converting-enzyme assay. J Comput Assist Tomogr 1998,22,229–34.

[58] Kohn H, Klech H, Mostbeck A, Kummer F. [67]Ga scanning for assessment of disease activity and therapy decisions in pulmonary sarcoidosis in comparison to chest radiography, serum ACE and blood T-lymphocytes. Eur J Nuc Med 1982,7,413–6.

[59] Bradley B, Branley HM, Egan JJ, Greaves MS, Hansell DM, Harrison NK, et al. Interstitial lung disease guideline: the British Thoracic Society in collaboration with the Thoracic Society of Australia and New Zealand and the Irish Thoracic Society. Thorax 2008,63 Suppl 5,v1–58.

[60] Dempsey OJ, Paterson EW, Kerr KM, Denison AR. Sarcoidosis. Br Med J 2009,339,b3206.

[61] Tahmoush AJ, Amir MS, Connor WW, Farry JK, Didato S, Ulhoa-Cintra A, et al. CSF-ACE activity in probable CNS neurosarcoidosis. Sarcoidosis Vasc Diffuse Lung Dis 2002,19,191–7.

[62] Izzo JL, Jr, Weir MR. Angiotensin-converting enzyme inhibitors. J Clin Hypertension (Greenwich) 2011,13,667–75.

[63] Pietila K, Koivula T. Increase of serum angiotensin-converting-enzyme activity after freezing. Scand J Clin Lab Invest 1984,44,453–5.

[64] Cushman DW, Cheung HS. Spectrophotometric assay and properties of the angiotensin-converting enzyme of rabbit lung. Biochem Pharmacol 1971,20,1637–48.

[65] Hayakari M, Kondo Y, Izumi H. A rapid and simple spectrophotometric assay of angiotensin-converting enzyme. Anal Biochem 1978,84,361–9.

[66] Hayakari M, Seito R, Furugori A, Hashimoto Y, Murakami S. An improved colorimetric assay of angiotensin-converting enzyme in serum. Clin Chim Acta 1984,144,71–5.

[67] Hurst PL, Lovell-Smith CJ. Optimized assay for serum angiotensin-converting enzyme activity. Clin Chem 1981,27,2048–52.

[68] Schnaith E, Beyrau R, Buckner B, Klein RM, Rick W. Optimized determination of angiotensin I-converting enzyme activity with hippuryl-L-histidyl-L-leucine as substrate. Clin Chim Acta 1994,227,145–58.

[69] Ryder KW, Jay SJ, Jackson SA, Hoke SR. Characterization of a spectrophotometric assay for angiotensin converting enzyme. Clin Chem 1981,27,530–4.
[70] Kasahara Y, Ashihara Y. Colorimetry of angiotensin-I converting enzyme activity in serum. Clin Chem 1981,27,1922–5.
[71] Peters RH, Golbach AJ, van den Bergh FA. Automated determination of angiotensin-converting enzyme in serum. Clin Chem 1987,33,1248–51.
[72] Filipovic N, Mijanovic M, Igic R. A simple spectrophotometric method for estimation of plasma angiotensin I converting enzyme activity. Clin Chim Acta 1978,88,173–5.
[73] Neels HM, van Sande ME, Scharpe SL. Sensitive colorimetric assay for angiotensin converting enzyme in serum. Clin Chem 1983,29,1399–403.
[74] Holmquist B, Bunning P, Riordan JF. A continuous spectrophotometric assay for angiotensin converting enzyme. Anal Biochem 1979,95,540–8.
[75] Gorski TP, Campbell DJ. Angiotensin-converting enzyme determination in plasma during therapy with converting enzyme inhibitor: two methods compared. Clin Chem 1991,37,1390–3.
[76] Wong YW, Kinniburgh DW. Evaluation of a kinetic assay for angiotensin converting enzyme. Clin Biochem 1987,20,323–7.
[77] Chiknas SG. A liquid chromatography-assisted assay for angiotensin-converting enzyme (peptidyl dipeptidase) in serum. Clin Chem 1979,25,1259–62.
[78] Horiuchi M, Fujimura K, Terashima T, Iso T. Method for determination of angiotensin-converting enzyme activity in blood and tissue by high-performance liquid chromatography. J Chromatogr 1982,233,123–30.
[79] Maurich V, Moneghini M, Pitotti A, Vio L. A simple high-performance liquid chromatographic method for estimating human serum angiotensin-converting enzyme activity. J Pharm Biomed Anal 1985,3,51–7.
[80] Maurich V, Pitotti A, Vio L, Mamolo MG. Ion-pair liquid chromatographic assay of angiotensin-converting enzyme activity. J Pharm Biomed Anal 1985,3,425–32.
[81] Meng QC, Berecek KH. Quantification of angiotensin-converting enzyme (ACE) activity. Methods Mol Med 2001,51,257–66.
[82] Shihabi ZK, Scaro J. Liquid-chromatographic assay of angiotensin-converting enzyme in serum. Clin Chem 1981,27,1669–71.
[83] Doig MT, Smiley JW. Direct injection assay of angiotensin-converting enzyme by high-performance liquid chromatography using a shielded hydrophobic phase column. J Chromatogr 1993,613,145–9.
[84] Neels HM, Scharpe SL, van Sande ME, Verkerk RM, Van Acker KJ. Improved micromethod for assay of serum angiotensin converting enzyme. Clin Chem 1982,28,1352–5.
[85] Badminton MN, Dawson CM, Rainbow SJ, Tickner TR. Use of a non-extraction HPLC technique for measuring angiotensin-converting enzyme under optimum conditions. Ann Clin Biochem 1991,28,396–400.
[86] Meng QC, Balcells E, Dell'Italia L, Durand J, Oparil S. Sensitive method for quantitation of angiotensin-converting enzyme (ACE) activity in tissue. Biochem Pharmacol 1995,50,1445–50.
[87] Friedland J, Silverstein E. A sensitive fluorimetric assay for serum angiotensin-converting enzyme. Am J Clin Pathol 1976,66,416–24.
[88] Friedland J, Silverstein E. Sensitive fluorimetric assay for serum angiotensin-converting enzyme with the natural substrate angiotensin I. Am J Clin Pathol 1977,68,225–8.
[89] Carmel A, Ehrlich-Rogozinsky S, Yaron A. A fluorimetric assay for angiotensin-I converting enzyme in human serum. Clin Chim Acta 1979,93,215–20.
[90] Maguire GA, Price CP. A kinetic fluorimetric assay for the measurement of angiotensin-converting enzyme in human serum. Ann Clin Biochem 1984,21 (Pt 5),372–7.
[91] Kapiloff MS, Strittmatter SM, Fricker LD, Snyder SH. A fluorometric assay for angiotensin-converting enzyme activity. Anal Biochem 1984,140,293–302.

[92] Ryan JW, Chung A, Ammons C, Carlton ML. A simple radioassay for angiotensin-converting enzyme. Biochem J 1977,167,501–4.

[93] Ryder KW, Thompson H, Smith D, Sample M, Sample RB, Oei TO. A radioassay for angiotensin converting enzyme. Clin Biochem 1984,17,302–5.

[94] Rohrbach MS. [Glycine-1-14c]hippuryl-L-histidyl-L-leucine: a substrate for the radiochemical assay of angiotensin converting enzyme. Anal Biochem 1978,84,272–6.

[95] Rohrbach MS, Deremee RA. Serum angiotensin converting enzyme activity in sarcoidosis as measured by a simple radiochemical assay. Am Rev Respir Dis 1979,119,761–7.

[96] Groff JL, Harp JB, DiGirolamo M. Simplified enzymatic assay of angiotensin-converting enzyme in serum. Clin Chem 1993,39,400–4.

[97] Zhang R, Xu X, Chen T, Li L, Rao P. An assay for angiotensin-converting enzyme using capillary zone electrophoresis. Anal Biochem 2000,280,286–90.

2 Acetylcholinesterase and butyrylcholinesterase

Peter L. Platteborze

2.1 Case studies

2.1.1 Patient A

A panicked mass of people arrived at the emergency department approximately 40 min after an alleged terrorist attack. The patients report a wide range of symptoms of varying severity that include miosis, rhinorrhea, blurry vision, nausea, broncho-constriction, dyspnea, and drooling. Serum samples were collected from all patients and analyzed immediately in the laboratory. All patients with symptoms were treated with atropine sulfate, and those with severe miosis and dyspnea were also given pralidoxime. Laboratory results were generally normal except several patients with the most severe complaints had undetectable serum enzyme butyrylcholinesterase activity (BChE, also called pseudocholinesterase; reference range: 4,900–11,900 U/L).

2.1.1.1 Discussion

The symptoms presented by the patients are consistent with inhalational exposure to a volatile anticholinesterase agent, possibly an organophosphorus nerve agent. These nerve agents are potent inhibitors of both acetylcholinesterase (AChE) and BChE. Inhibition of neuronal AChE results in the accumulation of the neurotransmitter acetylcholine at synapses which manifests in a cholinergic toxidrome with the signs and symptoms described above. Organophosphorus nerve agents remain a major chemical warfare threat. After exposure to an organophosphate, the serum BChE activity rapidly decreases and in severe cases can become undetectable. Both the type and timing of the treatments used were appropriate in these patients, as the degree of cholinesterase inhibition does not necessarily correlate with the degree of poisoning. Hence, diagnosis relies more upon the clinical presentation than laboratory results. Atropine competitively binds to muscarinic cholinergic receptors, thereby blocking the effects of excess acetylcholine and dramatically reducing most symptoms. Many organophosphates can irreversibly inhibit AChE due to a biochemical reaction known as "aging". The pralidoxime treatment serves to reactivate inhibited AChE prior to aging. If left untreated, the recovery of serum BChE and AChE activity will depend upon the production of a new enzyme. Untreated survivors with severe poisoning will typically recover normal BChE in approximately 50 days and AChE in approximately 120 days (0.8% per day). Subsequent laboratory analysis by gas chromatography-mass spectrometry confirmed that the liquid released was a crude form of the organophosphate sarin; metabolites of sarin were also identified in urine samples collected from several patients. Care of patients in attacks of this nature requires adequate precaution to be taken by caregivers. In the well-described Japanese subway incident, for example, many medical personnel treating the victims became contaminated with sarin and subsequently developed mild symptoms of poisoning.

2.1.2 Patient B

A 10-year-old female is seen in the emergency department because of a complaint of nausea, general muscle weakness and fasciculations, abdominal cramps, diarrhea, blurred vision, drooling, tearing, diaphoresis, dysarthria, and paresthesia of the lips and tongue. Her mother commented that she ate nearly an entire cantaloupe approximately 1 h prior to becoming ill and that the child had also experienced nausea. The attending physician diagnosed cholinesterase inhibitor poisoning due to the rapid appearance of symptoms. No other family members presented with symptoms, and none had consumed any cantaloupe. Blood samples were analyzed in the laboratory and showed a very low red blood cell (RBC) AChE (reference range: 25–52 U/g of hemoglobin). Emesis was induced, and the patient was treated with atropine sulfate, after which her symptoms rapidly resolved. Several hours later a new blood sample revealed that the AChE activity had increased significantly.

2.1.2.1 Discussion
The form of AChE that is anchored to the outer surface of RBC membranes can serve as an effective surrogate marker for neuronal AChE. Many argue that its activity better represents the degree of neuronal inhibition than the serum BChE test. Subsequent analysis of the remaining cantaloupe showed that it was contaminated with the carbamate pesticide aldicarb. Unlike many organophosphates, carbamate inhibition of AChE is rapidly reversible and aging cannot occur. The clinician suspected that this case involved inadvertent pesticide contamination. After receiving the laboratory results the poison center was contacted to report the case, which led to enhanced physician vigilance to monitor for food-borne anticholinesterase intoxication. Aldicarb is the active substance in the pesticide Temik, which the US Environmental Protection Agency has placed in its highest acute toxicity category. Diagnosis of intoxication is primarily clinical because the depression of cholinesterase activities is unreliable as an indicator of exposure. Multiple outbreaks of food-borne pesticide poisoning have been reported in the US that involved contaminated cantaloupes, cucumbers, watermelons, and mint. In addition, there have been reports of aldicarb poisoning due to its accidental addition to food, and it has been used in suicides. In the US military and several states in the US, RBC-AChE activities are used to monitor occupational exposure, usually by comparing the results from periodic assessments to known baseline values.

2.1.3 Patient C

A 60-year-old White male with aggressive prostate cancer was admitted to the hospital for a prostatectomy. He was healthy, did not consume alcohol, took a daily statin medication, and had no prior experience with general anesthesia. During surgery the short-acting muscle relaxant succinylcholine was administered to facilitate endotracheal intubation. The surgery was successful; however, the patient was unable to breathe independently 5 min after the removal of succinylcholine and thus was kept on a mechanical ventilator. The patient spontaneously recovered 2 h later with no apparent complications. An immediate post-operative blood specimen showed a normal basic metabolic panel and albumin levels; however, the serum BChE activity was 1,500 U/L (reference range: 4,900–11,900 U/L). The clinical team suspected that he had an inherited serum BChE deficiency and requested further confirmatory laboratory analysis. To elucidate the etiology of his apnea, another serum sample was drawn 48 h later and sent to a reference laboratory for dibucaine inhibition analysis. Unlike the wild-type BChE, a common BChE variant has been shown to be relatively resistant to inhibition by dibucaine.

2.1.3.1 Discussion

Serum BChE is responsible for metabolizing the commonly used surgical muscle relaxants succinylcholine (e.g., suxamethonium) and mivacurium. In patients with certain BChE mutations the degradation of these agents is impaired, with resulting prolonged paralysis and apnea after ceasing drug administration. The standard of care is to let the individual recover spontaneously. BChE deficiency can also be an acquired condition, often associated with liver disease. In these cases, there can be significantly decreased BChE activity that mimics genetic deficiency, leading to similar clinical symptomology. This patient's dibucaine inhibition test revealed a dibucaine number of 18. This number represents the percent of BChE activity inhibited by dibucaine. Patients with normal BChE alleles will have over 70% inhibition (dibucaine number >70), whereas individuals homozygous for the atypical BChE mutation will have less than 20% inhibition (dibucaine number <20). A dibucaine number between 40 and 70 is intermediate and is often due to heterozygosity. Individuals with two atypical BChE alleles would be expected to have prolonged post-succinylcholine induced apnea and paralysis. This patient had the atypical phenotype, as evidenced by low total BChE activity and low dibucaine number, which in this case caused the post-anesthesia complication. In cases like this, it is critical to wait at least 24 h after the final dose of succinylcholine prior to assaying with dibucaine. Because BChE mutations are moderately prevalent, genetic counseling of the patient and family members could be indicated.

2.1.4 Patient D

An approximately 25-year-old Hispanic male farm worker is seen in the emergency department. He complains of generalized muscle weakness, severe headaches, impaired vision, and copious diarrhea and vomiting. He also has shortness of breath, for which he was provided supplemental oxygen by face mask. The symptoms appeared 3 weeks earlier when he began spraying an apple orchard with insecticides, and they became progressively worse. On examination he appears to be tachypneic, diaphoretic, cyanotic, and he cannot stand up without assistance. An electrocardiogram shows a prolonged QTc interval and a portable chest radiograph shows pulmonary edema. After this testing he became comatose, was intubated, and repeated doses of atropine sulfate were administered until the majority of symptoms subsided. An infusion of pralidoxime was also given. A complete blood count and a basic metabolic chemistry panel were normal as was a rapid urine drug screen for drugs of abuse; the only abnormality found was a serum BChE activity of 300 U/L (reference range: 4,900–11,900 U/L). A phone call to his employer revealed that he had been working unsupervised 24 h earlier with chlorpyrifos. The next day he began regaining muscle strength and was extubated; his electrocardiogram returned to normal and the chest X-ray showed limited edema. He spent 4 days in the hospital during which time he slowly regained his strength. Three weeks after admission he returned to work; at that time his BChE was 5,000 U/L. Eight weeks post-admission his BChE was 10,000 U/L.

2.1.4.1 Discussion
The patient developed chronic occupational poisoning from using the organophosphate insecticide spray chlorpyrifos. Upon recovery, he revealed that due to the outdoor heat, he did not wear the personal protective equipment that was issued to him. Organophosphate insecticides can be potent inhibitors of both AChE and BChE, and are a major global health problem with an estimated 3 million poisonings annually. The treatment offered by the clinicians here was appropriate. Although, much of this patient's AChE had undergone biochemical "aging" due to chronic exposure, this was unknown at admission. Pralidoxime would likely have reactivated whatever AChE that had been inhibited by recent chlorpyrifos exposure, as aging takes approximately 24 h with this insecticide. It is worth noting that in humans, some pesticides can preferentially inhibit the serum BChE whereas others initially inhibit the RBC-AChE. Chronic chlorpyrifos exposure can cause cumulative AChE inhibition because repeated exposures close in time can reduce AChE faster than it can be regenerated.

2.1.5 Patient E

A 40-year-old African-American female with type 2 diabetes in the 17th week of her first pregnancy is seen to address concerns of a possible fetal abnormality. Her concern is that at approximately this same age, her mother delivered a stillborn child with spina bifida. Owing to her age, comorbidities, and family history, her pregnancy is being carefully monitored. The maternal serum α-fetoprotein (AFP) level was at seven multiples of the median, thus an amniocentesis and ultrasonography were performed. The amniotic fluid had an elevated AFP concentration, thus the specimen was reflexed to qualitative AChE testing by polyacrylamide gel electrophoresis. AChE was present in the amniotic fluid. AChE in amniotic fluid, together with significantly elevated AFP, is considered diagnostic for a fetal neural tube defect such as spina bifida. Ultrasonography confirmed the gestational age but resolution was insufficient to visualize an open tube neural defect. A therapeutic abortion was performed and the subsequent fetal autopsy confirmed spina bifida.

2.1.5.1 Discussion

Determination of AFP in amniotic fluid permits the diagnosis of most cases of spina bifida between 14 and 24 weeks of gestation. However, false-positive AFP results occur and are usually caused by blood contamination in the specimen. Furthermore, there is some overlap between the reference ranges of unaffected gestations and fetuses with neural tube defects, thus caution is necessary. Amniotic fluid AChE is the most specific biochemical marker for the presence of an open neural tube defect. Normally, only BChE is present in the fetal amniotic fluid; however, in an open neural tube defect, AChE leaks into the amniotic fluid from exposed fetal nerves. To detect AChE in amniotic fluid, proteins are first separated by polyacrylamide gel electrophoresis and then incubated with an AChE substrate such as acetylthiocholine iodide. If positive, two distinctive colored precipitant bands will appear illustrating cholinesterase activity, corresponding to the slower migrating BChE and the faster migrating AChE. The latter is confirmed by incubating another part of the gel with a specific AChE inhibitor. False-positives can occur if maternal or fetal hemoglobin is present in the amniotic fluid.

2.2 Biochemistry and physiology of the cholinesterases

2.2.1 Molecular forms

Humans express two different but related cholinesterase enzymes, AChE and BChE. These two homologous serine hydrolases have some overlapping substrate specificity because they both can hydrolyze acetylcholine. However, they can also be distinguished by substrate preference, inhibitor specificity, and antibody recognition. AChE (EC 3.1.1.7, acetylcholine acetylhydrolase) activity is critical to promptly hydrolyze the released acetylcholine at cholinergic nerve endings that mediates transmission of the neural impulses across the synapse. Common alternative names include erythrocyte or RBC cholinesterase and true cholinesterase; more are listed in Table 2.1.

BChE (EC 3.1.1.8, acylcholine acylhydrolase) was so named due to its ability to hydrolyze butyrylcholine faster than the many other esters it can degrade. It is abundantly expressed in the liver where it is efficiently secreted into the serum. It is also widely distributed throughout the human nervous system and appears to be expressed to some degree in most tissues with the exception of RBCs. Common alternative names are pseudocholinesterase and serum or plasma cholinesterase, additional names are listed in Table 2.1. Relative to AChE, BChE has unique enzymatic properties. Although it currently has no known physiological function, it is generally thought to be an important scavenger of naturally occurring and synthetic anticholinesterases. Recent studies suggest that it might also have functions in the normal nervous system. Laboratory based assays of both enzymes are clinically useful. A comprehensive review of these enzymes is in the published Proceedings of the Tenth International Cholinesterase Meeting held in 2009 [1].

2.2.1.1 AChE

AChE exists in multiple forms, all of which possess similar catalytic properties, but differ in their oligomeric assembly and mode of attachment to the cell surface. AChE is among the fastest known enzymes with the hydrolysis of acetylcholine to acetate

Table 2.1: General background of the human cholinesterases.

	Acetylcholinesterase (EC 3.1.1.7)[1]	Butyrylcholinesterase (EC 3.1.1.8)
Systematic name	Acetylcholine acetylhydrolase	Acylcholine acylhydrolase
Synonyms	Erythrocyte cholinesterase	Pseudocholinesterase
	Red blood cell (RBC) cholinesterase	Serum or plasma cholinesterase
	True cholinesterase	False cholinesterase
	Specific cholinesterase	Non-specific cholinesterase
	Choline esterase I	Choline esterase II
		Benzoyl cholinesterase
Abbreviation	AChE	BChE
Major tissue presence	CNS, PNS, RBCs, lung, skeletal muscle, spleen, gray matter of the brain	Blood, liver, CNS, PNS, heart, pancreas, white matter of the brain
Function	Hydrolysis of acetylcholine	Undetermined
Substrate preference	Hydrolysis of acetyl esters	Hydrolysis of various choline esters
Optimum substrate	Acetylcholine	Butyrylcholine

Note: 1. AChE is expressed primarily at neuromuscular junctions in the peripheral nervous system (PNS) and cholinergic synapses in the central nervous system (CNS). It is also found in the lungs, spleen, the gray matter of the brain, and an alternative form is anchored onto the surface of RBC membranes where it constitutes the YT blood group antigen.

and choline approaching the maximal theoretical limit set by molecular diffusion of the substrate, see Equation (2.1).

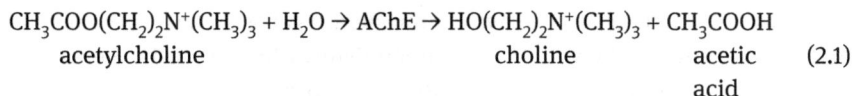

$$CH_3COO(CH_2)_2N^+(CH_3)_3 + H_2O \rightarrow AChE \rightarrow HO(CH_2)_2N^+(CH_3)_3 + CH_3COOH$$

acetylcholine   choline   acetic   (2.1)

acid

This rapid reaction is integral to the prompt termination of neurotransmission at cholinergic synapses. Based upon the impressively high turnover number, the crystallographic structure of AChE was totally unexpected. The active site is buried at the bottom of a deep narrow gorge which should restrict substrate accessibility. Additionally, the catalytic gorge is lined by aromatic residues and lacks negatively charged residues that were theoretically presumed important to interact with the quaternary nitrogen in acetylcholine. The catalytic binding site contains the catalytic triad of Ser200, His440, and Glu327. Briefly, the catalytic mechanism of AChE involves an initial acylation step where the active site serine reacts with the acetylcholine causing the choline to be displaced and forming an acylated serine. This choline then diffuses away. This biochemical reaction is facilitated by other strategically proximal residues, especially the active site glutamate, that orient the positively charged acetylcholine substrate to the best location for serine to displace the choline. In the next step, water bound to the active site histidine attacks the acyl group displacing it from the serine and generating acetic acid. This molecule diffuses away leaving a regenerated AChE that can rapidly repeat this reaction.

The *AChE* gene is located on chromosome 7 at 7q22 and spans approximately 7 kb of genome. It also corresponds to the YT1 blood group antigen and a His322Asn mutation causes the rare YT2 blood group. Three invariant exons encode the signal sequence and the amino terminal 543 amino acids common to all forms of the enzyme. Alternative exon splicing of the next exon accounts for the structural divergence in the carboxyl termini of catalytic subunits. These structural differences dictate the cellular disposition of the three different AChE forms but do not influence catalytic activity. These three species include a form associated with the specific collagen ColQ, another form associated with the transmembrane protein PRiMA, and a form anchored on the surface of RBC membranes by the post-translational addition of a glycophosphatidyliositol. Because obtaining AChE from cholinergic synapses is invasive and impractical for routine testing purposes, RBC-AChE is commonly sampled as a surrogate indicator of neuronal AChE inhibition.

Each AChE monomer is glycosylated at three asparagines and contains three internal disulfide bonds plus an odd cysteine that forms a disulfide linkage to an identical catalytic subunit to form a dimer [2]. These dimers often non-covalently associate to form homomeric tetramers. Heteromeric oligomers can also form between the catalytic subunits and either ColQ or PRiMA. The former is usually expressed at neuromuscular junctions whereas the latter is expressed on the surface of neurons. Very little free AChE is present in plasma, on average less than 8 ng/mL.

2.2.1.2 BChE

BChE is structurally and functionally related to AChE. Both belong to the α-β hydrolase fold family of proteins as they contain a central β-sheet that is surrounded by α-helices. It is thought that the two cholinesterase genes arose from gene duplication at the emergence of the vertebrates. Functionally, BChE can catalyze the hydrolysis of acetylcholine, although less efficiently than AChE. It also hydrolyzes other choline esters, such as those found in succinylcholine (e.g., suxamethonium), mivacurium, procaine, chloroprocaine, tetracaine, cocaine, and heroin. Significantly reduced BChE activity can result in the prolongation of succinylcholine or mivacurium induced neuromuscular blockade. BChE deficiency can be inherited or acquired; the latter can involve a number of conditions. Membrane-bound BChE exists in the brain, peripheral nerves, and muscles. Based upon the critical role of AChE, it was very surprising that AChE knockout mice could survive although severely compromised and with a shortened lifespan. This suggests that the BChE could substitute for AChE to hydrolyze acetylcholine [3]. It is worth noting that apparently normal, healthy people with no detectable serum BChE activity seem to be at no disadvantage.

The *BChE* gene is located on chromosome 3 at 3q26.1–26.2 and contains four exons spanning over 70 kb of genome. The enzyme is highly expressed by the liver and immediately secreted into circulation. The average adult human plasma contains over 3 µg/mL of BChE as measured by immunoassay. Although the true function of BChE has not been determined, it has been shown to be important in scavenging naturally occurring anticholinesterases such as physostigmine and synthetic anticholinesterase such as organophosphate pesticides. Similar to AChE, the enzyme active site contains a catalytic triad comprising Ser226, His466, and Glu353. There are 574 amino acids in the mature protein and each monomer contains three internal disulfide bonds, one interchain disulfide bond, and nine asparagine-linked carbohydrate chains. Circulating BChE exists as a tetramer of four identical subunits and has a relative molecular mass of 342 kDa, approximately 24% of which is due to glycosylation [4]. Serum BChE has an estimated half-life of 10 to 14 days. Generally, exposure to anticholinesterase compounds results in BChE activity declining much more rapidly than RBC-AChE. Because of this, serum BChE testing has become a routine laboratory assay to monitor acute poisoning.

2.2.2 Inheritance of BChE variants

Not long after the short-acting muscle relaxant succinylcholine was introduced in 1951, it was observed that patients deficient in BChE experienced a prolonged apnea that required continued artificial respiratory support. This occasionally resulted in deaths. It was subsequently determined that many of these patients had an unusual form of BChE that only slowly degraded succinylcholine. Additional characterization revealed that unlike the wild-type BChE, this atypical inherited form also resisted *in vitro* inhibition by dibucaine. The discovery of this inherited dibucaine-resistant BChE was one of the first applications of pharmacogenetics and it led to the discovery of the atypical BChE allele. Currently, there are at least 65 different autosomal recessive variants that can result in minimal to extremely dangerous post-succinylcholine apnea and paralysis [5]. It is estimated that 5% of the population may have abnormal inherited forms. The most commonly discussed BChE variants are the atypical (A) or dibucaine-resistant, fluoride-resistant (F), silent (S), Kalow (K), and James (J) forms. The wild-type form of BChE is commonly referred to as the usual (U) form. A summary of these common BChE variants is presented in Table 2.2.

Table 2.2: Common BChE variants[1].

BChE phenotype	Mutation	Estimated frequency[2]	Sensitivity to succinylcholine	Estimated apnea	BChE activity
Usual (U)	Wild type	Approximately 95% of the general population	UU – normal response	Normal: 3–5 min	UU – normal 100%
Atypical (A) dibucaine-resistant	D70G	Homozygous: 0.03–0.01%	AA – very sensitive	≥2 h	AA – reduced 70%
Fluoride-resistant (F)	F-1:T243M F-2:G390V	Heterozygous: 4% Homozygous: 0.07% Heterozygous more common and less clinically relevant	FF – moderately sensitive	1–2 h	FF – reduced 60%
Kalow (K)	A539T	Homozygous: 1.5%: often associated with other variants	KK – mildly sensitive	<1 h	KK – reduced 30%
James (J)	E497V	Homozygous: 0.0007%: often associated with other variants	JJ – moderately sensitive	1–2 h	JJ – reduced 67%
Silent (S)	Over 20	Homozygous: 0.01–0.008% Heterozygous: 0.7% but less clinically relevant	SS – extremely sensitive	≥3 h	SS – reduced >98%

Notes: 1. Variants can exist by themselves or with others making their response more variable than indicated. 2. Frequency estimates for the general population; shown to vary among different populations.

All of these variants except some of the S forms are due to single nucleotide polymorphisms. Many of the S forms are due to frame shift mutations, deletions or insertions, that can generate a premature stop codon. Whereas only one single nucleotide mutation has been found to cause the A, K, and J phenotypes, two different mutations can cause the F type, and at least 20 are found to result in S variants. The effect of these BChE mutations is to either alter the ability of the enzymes to efficiently hydrolyze substrate or to reduce the amount of gene product and thus enzyme concentration in blood. The former mutations are called qualitative variants and are represented by A, F, and some S phenotypes. The latter mutations lead to quantitative variants and are represented by K, J, and many S phenotypes. Relative to the usual BChE, the K variant has approximately 33% lower activity, the J variant has 67% lower activity, and the S variants have more than 98% lower activity. Both qualitative and quantitative mutations can reduce the BChE activity determined in the laboratory. Extremely rare variants have been identified that have greater BChE activity than the usual form and these individuals can be resistant to a normal dose of succinylcholine.

The atypical (A) allele is relatively common and the phenotype is determined by dibucaine resistance [6]. The percent of BChE activity inhibited by the addition of dibucaine is called the dibucaine number. A dibucaine number above 70 describes typical or normal BChE, a number between 40 and 70 is intermediate and is often associated with (A) heterozygotes, and a number below 20 is atypical and associated with (AA) homozygotes. Patients with a normal dibucaine number would react normally to succinylcholine, BChE (A) heterozygotes would generally be expected to have a normal or minimally prolonged response to succinylcholine, and (AA) homozygotes would be expected to have a prolonged post-succinylcholine apnea lasting 2 h or more. Similar prolonged apnea related to BChE also occurs after the administration of the muscle relaxant mivacurium. Homozygotic A variants can also show significant side effects after the ingestion of the anticholinesterase pyridostigmine. Estimates of the frequencies of common BChE variants are listed in Table 2.2. Some of these frequencies have been shown to vary by ethnic populations. For example, the A allele is almost absent in the Japanese yet is common in populations from areas where plants of the Solanaceae family, such as potatoes and tomatoes, originated. This distribution is presumed due to the wild-type BChE being inhibited by solanaceous glycoalkaloids, whereas the A form is resistant and possibly providing an evolutionary advantage. Occasionally, BChE genetic variants are not realized until these individuals are exposed to succinylcholine or mivacurium as illustrated earlier in patient C who after surgery was identified with an atypical dibucaine-resistant phenotype and is probably homozygous for A. Individuals homozygous for S exposed to succinylcholine or mivacurium will have an even more profound paralysis and apnea.

Subsequent to the discovery of dibucaine-resistant mutants, fluoride-resistant variants were identified. The F variants do not necessarily show resistance to inhibition by dibucaine, thus fluoride numbers was added as another analytical tool for identifying BChE variants. A normal fluoride number is from 55 to 65 and homozygotes would be expected to have a moderate sensitivity to succinylcholine and mivacurium.

In addition to homozygotes, all combinations of heterozygotes have been found and it is not rare for multiple mutations to occur within a single *BChE* gene or for there to be a combination of heterozygotic states. In North America, approximately 90% of people with the A variant will also have the K mutation on the same chromosome. Similarly, the J phenotype is often associated with the K mutation. The common K variant alone is of minimal clinical significance; however, in combination with other BChE mutations and/or the presence of underlying acquired deficiencies there can be clinically significant reductions of BChE activity. These combinations help to explain the wide range of BChE activities observed in the biochemical phenotypes. No biochemical test currently exists that can detect all BChE variants. Traditional assays such as the dibucaine and fluoride inhibition tests are not sufficient to identify all the currently identified variants and some instances of prolonged apnea due to succinylcholine and mivacurium remain unexplained. Because BChE is a small gene containing only four exons, DNA testing for the deficiency seems feasible.

2.3 Chemical pathology

Laboratory based measurements of the cholinesterases have many diverse clinical applications. Foremost is for the detection of possible insecticide or chemical warfare nerve agent poisoning. These compounds are highly toxic owing to their ability to inhibit AChE at neurons. Both serum BChE and RBC-AChE serve as effective surrogate markers. Serum BChE activity measurements are also used to detect patients with gene variants that can result in life-threatening sensitivity to the commonly used surgical drugs succinylcholine and mivacurium. BChE measurements have also been used to evaluate liver function. Testing of amniotic fluid for the presence of fetal AChE can be used to diagnose open neural tube defects. In addition, anticholinesterase medications have been used to treat several disease conditions.

2.3.1 Pesticide and nerve agent poisoning

Both cholinesterase enzymes are selectively inhibited by organophosphorus compounds and N-methyl carbamates. The degree of serum BChE and RBC-AChE inhibition is a reflection of the degree of neuronal inhibition. As such, their activities are often measured in the laboratory to assess potential human exposures to these abundantly used compounds. To date, more than 200 organophosphates and 25 N-methyl carbamates have been formulated into thousands of different products used worldwide. Occasionally these are used medicinally, but most are toxic pesticides that are generally well absorbed in the lungs, gastrointestinal tract, skin, and mucus membranes of exposed human following inhalation, ingestion, or topical contact. The World Health Organization estimates that there are at least 3 million global poisonings annually from these pesticides: 1 million unintentional poisonings and 2 million suicide attempts. From this there are at least 200,000 annual fatalities. This toxicological poisoning problem is less pervasive in the USA, possibly owing to better public education and governmental oversight, as well as the use of less toxic pesticides. Records from the American Association of Poison Control Centers indicate that over a recent 5-year period (1998–2002) there were over 55,000 exposures to organophosphates and more than 25,000 exposures to carbamates [7]. From this there were on average eight annual fatalities. This public problem extends beyond pesticide exposures as chemical warfare nerve agents have been utilized on the modern battlefield and by terrorists. The two Japanese sarin incidents that occurred in the mid-1990s resulted in 1,200 victims with 17 deaths. Some common organophosphates include chlorpyrifos, parathion, malathion, diazinon, dursban, chemical warfare nerve agents (tabun, sarin, soman, and VX), and the medicinal drugs echothiophate, and tacrine. Some common carbamates include aldicarb, carbaryl (sevin), carbofuran (furadan), and the medicinal compounds pyridostigmine, physostigmine, neostigmine, and donepezil.

The specific inhibition of the cholinesterases is the consequence of carbamylation (by the carbamates) or phosphorylation (by the organophosphates) of the active site serine hydroxyl group. In carbamate poisoning, the carbamyl-serine bond spontaneously hydrolyzes, thereby reactivating the enzyme. This occurs approximately 1 h after reaction with physostigmine and from 4 to 6 h with pyridostigmine. Owing to this spontaneous reactivation and the general inability to penetrate the CNS, carbamate toxicity is less severe and of shorter duration than that for the organophosphates. The duration of symptoms in carbamate insecticide poisoning is usually less than 24 h. In organophosphate poisoning, the alkylphosphoryl-serine bond also undergoes spontaneous hydrolysis with enzyme reactivation; however, the rate is very slow. The stability of this chemical bond depends on the size of the alkyl groups; hydrolysis occurs for small alkyl groups (e.g., methyl or ethyl) but for the larger ones cleavage may not occur. Instead, these large phosphorylated enzyme complexes lose an alkyl group at variable rates by hydroxylation to form a highly stable organophosphoryl-serine bond that is fully resistant to even pharmacologically mediated hydrolysis. This dealkylation reaction, termed enzyme "aging", causes irreversible inhibition and cholinesterase activity returns only with the synthesis of a new enzyme. "Aging" half-times vary between the organophosphates and range from 2 to 6 min with soman, to 3 to 5 h with sarin, and from 24 to 48 h with VX and most insecticides [8].

Both organophosphorus and carbamate poisonings are usually amenable to therapy. Therapeutic intervention commonly involves administration of atropine and if the poisoning is severe, pralidoxime is also given. Atropine acts on the muscarinic acetylcholine receptors at neuronal synapses to decrease their sensitivity to the accumulated acetylcholine; this consequently ameliorates many of the effects of the cholinesterase inhibitors. This includes the effects on the respiratory, gastrointestinal, and cardiac functions, but atropine has no effect at the nicotinic receptors of skeletal muscles. Pralidoxime is used to reactivate inhibited AChE. It binds to the active site of the enzyme and the nucleophilic attack by its oxime group causes the dephosphorylation or decarbamylation of the serine group. Timely treatment is critical as pralidoxime is unable to reactivate the "aged" form of the phosphorylated cholinesterases. The cases presented in patients A, B, and D illustrated the appropriate uses of these therapies.

The major consequence of AChE inhibition is that increased concentrations of acetylcholine accumulate at cholinergic synaptic junctions. This excess neurotransmitter stimulates cholinergic muscarinic receptors (PNS and CNS) and stimulates but then depresses or paralyzes cholinergic nicotinic receptors. Where this occurs, the membrane of the post-synaptic cell repetitiously depolarizes, producing action potentials in post-synaptic neurons and muscle contraction in post-synaptic skeletal and smooth muscle. In these muscles, the intracellular calcium released during each depolarization accumulates to such a level that prevents the muscle from relaxing. In skeletal muscle this clinically manifests as muscle fasciculations leading to paralysis. The effects on the smooth muscle of the gastrointestinal and respiratory tracts due to overstimulation of muscarinic receptors are also observable. Prominent effects include excessive salivation, nausea, vomiting, diarrhea, abdominal cramping, and asthmatic symptoms due to a combination of bronchoconstriction and bronchorrhea. Similar effects at other muscarinic sites produce excessive sweating, potentially painful miosis, and blurry vision. Effects on the cardiovascular system can produce life-threatening bradycardia, arrhythmias, and pulmonary edema. Overstimulation of CNS muscarinic receptors can cause ataxia, confusion, slurred speech, seizures, coma, and eventually central respiratory paralysis. Common signs and symptoms observed in anticholinesterase poisonings are listed as follows:

- Abdominal cramps
- Agitation
- Ataxia
- Blurred vision
- Bradycardia
- Bronchorrhea
- Bronchoconstriction
- Circulatory collapse
- Confusion
- Defecation/diarrhea
- Dyspnea
- Diaphoresis
- Emesis
- Hypertension
- Insomnia

- Lacrimation
- Miosis
- Muscle fasciculations
- Muscle weakness
- Respiratory depression
- Rhinorrhea
- Salivation
- Seizures
- Tachycardia
- Tremors
- Urinary incontinence

The activation of peripheral muscarinic receptors causes classic signs and symptoms classified as the cholinergic crisis or cholinergic toxidrome. These are described by the mnemonics MDSLUDGE (miosis, diaphoresis, salivation, lacrimation, urination, defection, gastrointestinal distress, emesis) or DUMB BELS (diarrhea, urination, miosis, bradycardia, bronchorrhea-bronchoconstriction, emesis, lacrimation, salivation-sweating). The actual clinical presentation observed in poisoned patients depends on the balance between muscarinic and nicotinic receptor stimulation. Although miosis is most common, especially in vapor exposures as illustrated in patients A and D, it may not always occur and mydriasis may be present (nicotinic action). Similarly, tachycardia (nicotinic) may be observed rather than bradycardia (muscarinic). Fatalities are usually due to respiratory failure caused by nicotinic receptor mediated muscle paralysis, combined with muscarinic receptor mediated bronchorrhea, bronchoconstriction, and CNS depression. Death can occur as early as 5 min after a single acute exposure to an organophosphate nerve agent.

The inhibition of AChE at neuronal synapses cannot be measured directly, instead surrogate markers are measured. The activities of both serum BChE and RBC-AChE decrease appreciably following poison exposures and thus both are used to diagnose and monitor poisonings. Each analyte has its own merits for being tested in the clinical laboratory. Serum BChE measurements are often used to monitor for acute poisoning because its activity declines and returns to normal more rapidly than that for RBC-AChE. Furthermore, measurement of BChE activity is easier to perform than that for RBC-AChE, thus it is more likely to be performed in hospital clinical laboratories. However, BChE testing does have some major limitations. Reduced serum BChE activity has been observed in patients with a number of other conditions, such as hereditary deficiency, pregnancy, malnutrition, and hepatic diseases. Additionally, a spectrum of drugs has been shown to depress its activity. The observed intraindividual variability of BChE is significant: it is estimated to be as high as 20% in a healthy population. When interpreting BChE activities it is important to remember that the reference values for the enzyme are sufficiently wide such that some individuals could lose half of their activity and still be inside the normal reference range.

RBC-AChE activity is generally considered more specific than serum BChE to reflect the degree of *in vivo* neurotoxicity because it is catalytically similar to the inhibited neuronal enzymes. In addition, RBC-AChE has a narrower normal reference range than BChE which should permit better identification of poisoned individuals. After an organophosphate exposure RBC-AChE recovers more slowly than serum BChE, thus making it a more sensitive indicator of past (chronic) exposure. Workers in contact with organophosphates should have a baseline AChE level determined for comparison and monitoring. In the absence of this, recovery is best assessed by the RBC-AChE activity achieving a steady state and not simply by returning to within the reference range. There are some limitations with the RBC-AChE assay. Depressed RBC-AChE activity may be due to exposures or conditions other than insecticide poisoning, such as antimalarial therapy, pernicious anemia, and hemoglobinopathies. Furthermore, daily intra-individual variability can be as high as 10% and individuals taking oral contraceptives can have elevated activity. Field monitoring studies of orchard workers have illustrated that lower AChE levels were more indicative of chronic (cumulative) exposures, whereas BChE was more responsive to acute exposures. Although usually not practical, ideally both serum BChE and RBC-AChE should be measured to generate the best developed clinical picture. Research has shown that some pesticides have preferential affinities for either RBC-AChE or serum BChE. A summary of differences in the laboratory assays is shown in Table 2.3.

Using cholinesterase testing data to make medical decisions is inherently complicated. For clinical decision making that involves acute poisoning, the value of cholinesterase test data for the diagnosis, treatment, and management of patients is problematic. Assessments often depend on whether or not the poison is known or suspected to be an anticholinesterase, approximate time exposure occurred, and whether the patient has baseline or pre-exposure activity measurements. The importance of the latter is better understood when one considers the extent of inter- and

Table 2.3: Comparison of cholinesterase laboratory tests.

	RBC AChE	Serum BChE
Advantage	More accurate reflection of synaptic inhibition; narrower reference range	Easier to assay in the laboratory, more commercially available
Regeneration of activity if untreated	0.8% per day	25–30% in first 7–10 days
Onset of depressed enzyme activity	Late, activity declines more slowly than BChE	Early, activity declines rapidly
Common clinical utilization	Chronic exposures, less utility with acute exposures	Acute exposures
Non-chemical causes of depression	Pernicious anemia, hemoglobinopathies	Liver diseases, many gene variants, pregnancy, malnutrition

intra-individual variability of normal cholinesterase levels and, especially for BChE, the many other interferences that can cause decreased activity. The degree of inhibition often does not closely correlate with the clinical severity of poisoning. However, the general literature cites that general signs and symptoms of poisoning can begin when activity is inhibited by approximately 50% of the lower limits of normal and serious neuromuscular symptoms can manifest at decreases of approximately 80%. Therapeutic decisions for acute exposures are guided primarily by patient history and clinical presentation, not by either cholinesterase activity. In fact, by the time patients present with acute symptoms the levels of both cholinesterases have typically fallen significantly below baseline values, often below detectable limits as seen in patients A and D.

Cholinesterase testing is usually considered more valuable for decision making that involves medical surveillance and/or biological monitoring than for clinical diagnosis and treatment of acute poisoning. Here, the medical decision usually involves whether to remove employees from workplace exposures. In the USA, thousands of agricultural workers are routinely tested for reduced cholinesterase activity relative to baseline levels. General guidelines state the workplace should be evaluated if an employee's RBC-AChE or serum BChE is less than 80% of baseline. Workers are required to leave the workplace if their levels are below 70% for RBC-AChE and below 60% for BChE. They cannot return to work until their cholinesterase activity has recovered to 80% of baseline [9].

All cholinesterase testing should be consistently done at the same laboratory using the same method. Because neither the serum BChE nor RBC-AChE methods have been standardized as a rule, one should not directly compare cholinesterase laboratory results from different laboratories. Analysts should also be aware that for severe carbamate exposures, the cholinesterase laboratory assays may produce false normal or moderately reduced activities due to the spontaneous regeneration of the enzyme before or during the assay. The most definitive means of determining exposure is the measurement of specific urinary metabolites by gas chromatography-mass spectrometry. This is usually unnecessary and unavailable in a sufficiently timely manner to support the emergency management of acute poisonings.

2.3.1.1 Liver disease

BChE measurement can serve as an indicator of the synthetic capacity of the liver, its decrease generally parallels that of serum albumin levels. Decreases in BChE have been observed in hepatic diseases such as hepatitis, cirrhosis, and carcinomas with metastases to the liver. A 30% to 50% decrease in activity has been observed in acute hepatitis and in chronic hepatitis of extended duration [10]. Decreases from 50% to 70% occur in advanced cirrhosis and carcinomas, with hepatic cancers demonstrating the greatest reduction followed by lung, gastrointestinal, and genitourinary malignancies. BChE testing has also been used to monitor liver function after liver transplantation. In patients with end-stage liver disease, normal BChE levels were regained immediately after liver transplant. Serial BChE measurements have also been used as a prognostic indicator in patients with liver disease. Reduction of serum BChE to below 25% of a patient's pre-disease baseline level is associated with impending death. However, as the half-life of serum BChE is estimated at 10–14 days, severe fulminant liver disease can cause death prior to a substantial decrease in BChE. The specificity of BChE for these indications is further diminished due to the significant daily variation in patients (±20%) and many other factors can reduce BChE such as pregnancy, malnourishment, medications, gene variants, and possibly diet (e.g., consumption of solanaceous plants). In fact, late pregnancy can reduce BChE activity an estimated 30% [11]. The use of serum BChE measurement in liver disease varies between countries. It has seldom been used in North America because it is not considered cost effective or reliable but has historically been used in Germany, Italy, and Japan.

2.3.1.2 Other BChE

As discussed previously and illustrated in patient C, BChE has a critical role in hydrolyzing the short-acting muscle relaxants succinylcholine and mivacurium. Dibucaine and fluoride inhibition tests can distinguish inherited forms of BChE deficiency from acquired forms. BChE is also involved in the metabolism of other drugs, such as cocaine and heroin, and several local anesthetics such as procaine. Research has been conducted to increase the rate at which BChE can hydrolyze cocaine. The US military has investigated the use of BChE as a stoichiometric scavenger to prophylactically bind nerve agents prior to significant neuronal inhibition [8]. Site-directed mutagenesis has also been conducted to confer upon BChE the capability to hydrolyze nerve agents. Decreased BChE activity has been correlated with a range of other conditions not mentioned previously. These include chronic renal disease, burns, malabsorption, myocardial infarction, congestive heart failure, pulmonary edema, leprosy and acute infections, or after surgery.

2.3.1.3 Other AChE

As illustrated in patient E, the presence of fetal AChE in amniotic fluid can serve as a diagnostic marker for open neural tube defects. It is worth noting that select carbamate and organophosphate AChE inhibitors are used medicinally to treat a range of diseases. AChE inhibitors, such as tacrine and donepezil, are among the Food and Drug Administration (FDA)-approved drugs to manage Alzheimer's disease, which is characterized by extensive neurodegeneration. The therapeutic rationale is that the relatively selective loss of basal forebrain cholinergic neurons depletes cortical acetylcholine, thus the presence of anticholinesterases should increase and prolong the availability of acetylcholine [12]. This could result in improved cognitive functions, such as memory and attention. Myasthenia gravis is an autoimmune disease caused by antibodies that can bind to nicotinic cholinergic receptors at neuromuscular junctions causing them to be destroyed or inhibited. This culminates in general muscle weakness and fatigability. Anticholinesterase therapy improves muscle function by reducing AChE activity permitting acetylcholine more time to stimulate functioning receptors. At least 16,000 Americans use pyridostigmine daily for myasthenia gravis. The disease glaucoma is often associated with increased intraocular pressure that results in optic nerve damage that can lead to progressive irreversible loss of vision. Anticholinesterase medications, such as echothiophate eye drops, are therapeutically useful by acting as ocular hypertensives. Decreased AChE activity has also been correlated with a range of other conditions not mentioned previously. These include paroxysmal nocturnal hemoglobinuria, active pyridostigmine treatment, and the relapse of megaloblastic anemia.

2.4 Analytical measurements of AChE and BChE

2.4.1 Desired specimens

RBC-AChE analysis uses unfrozen whole blood; acceptable anticoagulants are heparin, EDTA, or acid-citrate-dextrose (ACD). Specimens are stable at 4°C for 20 days in ACD or 6 days in EDTA or heparin, at room temperature for 5 days in ACD and 2 days in EDTA or heparin [13]. Amniotic fluids must be unfrozen and free of hemolysis. For assaying BChE, serum is preferred but plasma anticoagulated with heparin or EDTA is acceptable. Oxalate, fluoride, and citrate need to be avoided because they inhibit BChE. Serum or plasma should be separated from cells as soon as possible and hemolysis avoided. Specimens are stable at room temperature for 6 h and 1 week at 4°C. The enzyme is very stable frozen, at –20°C plasma specimens retained more than 95% of their original activity after 3 years of storage. Activity can also be preserved in dried blood specimens for forensic analysis.

2.4.1.1 Methods for measurement of cholinesterase
The most commonly used commercial assays to measure either RBC-AChE or BChE activity utilizes the Ellman chromogenic method [14, 15]. This methodology is based upon thiocholine esters having a rapid rate of hydrolysis and is followed by a colorimetric reaction involving the active thiol group present on the liberated thiocholine. Large absorbance increases are observed from small sample volumes and the increase in absorbance is directly proportional to the amount of enzyme present in the sample. The current Ellman reaction typically uses acetylthiocholine or butyrylthiocholine substrate for the RBC-AChE and BChE assays, respectively. These substrates are enzymatically hydrolyzed to form thiocholine; the free sulfhydryl present in thiocholine

can then chemically react with dithiobis-2-nitrobenzoic acid [e.g., DTNB, Ellman reagent] to form the yellow colored product 5-thio-2-nitrobenzoic acid, see Equation (2.2). This is spectrophotometrically measured at a wavelength from 400 to 420 nm. The results are typically expressed as micromoles per minute per milliliter (e.g., units/ mL) and for AChE, units per gram of hemoglobin. This assay has been shown to be a simple, reliable, accurate, sensitive, and inexpensive procedure and many variations of the method have been published.

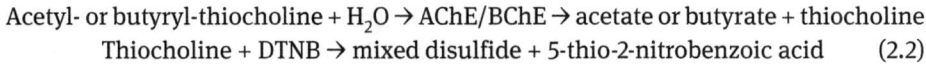

$$\text{Acetyl- or butyryl-thiocholine} + H_2O \rightarrow \text{AChE/BChE} \rightarrow \text{acetate or butyrate} + \text{thiocholine}$$
$$\text{Thiocholine} + \text{DTNB} \rightarrow \text{mixed disulfide} + \text{5-thio-2-nitrobenzoic acid} \qquad (2.2)$$

A range of conditions and substrates can be used in both cholinesterase assays, thus reference ranges vary between methods and laboratories. Based upon a 2010 College of American Pathology survey, the most commonly used substrate to measure BChE activity is butyrylthiocholine. Some other BChE substrates that are currently used in the Ellman reaction are acetylthiocholine, propionylthiocholine, and succinyldithiocholine [6]. Because AChE is anchored to RBCs, the activity obtained is dependent on the method used to lyse the RBCs and release the enzyme into solution. Procedures involving RBC-AChE must take into account that the lysing reagent can have inhibitory effects on activity, and corrections of absorbance for the RBC ghosts may be necessary.

Cholinesterase activity testing can also be done in the field using the Test-Mate System (EQM Research, Inc., Cincinnati, OH, USA) or the Lovibond Cholinesterase AF267 Test Kit (Tintometer Ltd, London, UK) [8]. The widely used Test-Mate System uses the Ellman method and a battery operated fixed-wavelength (450 nm) photometer. It requires 10 µL of blood and the test can be completed in less than 4 min. Either RBC-AChE or BChE can be measured depending on the assay kit with the former containing a specific BChE inhibitor. The Test-Mate System also measures hemoglobin, thus RBC-AChE results can be expressed as U/mL or U/g of hemoglobin. The Lovibond Cholinesterase AF267 Test System uses a subjective color comparator.

In addition to the Ellman chromogenic methods, there are two less commonly utilized cholinesterase assays [8, 9]. The delta pH method measures the change in acid levels over a specified time and laboratory values are usually reported as units of delta pH per hour. This test is based upon the addition of the substrate acetylcholine or butyrylcholine to the sample, which is subsequently hydrolyzed by the enzyme present to produce acetic or butyric acid and choline. The accumulation of these acid products results in an overall pH change that is measured with a pH meter. The delta pH method is slow and has a low throughput, which limits its commercial feasibility. However, the US Army's Cholinesterase Reference Laboratory annually performs over 15,000 delta pH AChE assays to support their occupational medical surveillance program. A radioactive acetylcholine/butyrylcholine microassay has also been used that measures radiolabeled products generated from labeled substrates following enzymatic reaction and an organic-water extraction. Although this method is very accurate, it is seldom used due to significant reagent expense and radioactive waste disposal issues.

As illustrated in patient E, amniotic fluid can be qualitatively examined for fetal AChE, which is a diagnostic indicator of open neural tube defects. To detect AChE in amniotic fluid, proteins are first separated by polyacrylamide gel electrophoresis and then incubated with an AChE substrate such as acetylthiocholine iodide. If positive, two distinct stained bands will appear illustrating cholinesterase activity, a slower migrating BChE and the faster migrating AChE. The AChE is confirmed by incubating another gel with an inhibitor specific to AChE, repeating the colorimetric reaction, and observing only the BChE band. False-positives can occur if maternal or fetal hemoglobin is present in the amniotic fluid.

As illustrated in patient C, some individuals have BChE mutations that make them sensitive to prolonged apnea after administration of succinylcholine or mivacurium. Laboratory determined dibucaine and fluoride numbers are useful to rule out an acquired deficiency and to phenotype homozygous and heterozygous patients. The Ellman method is used in the presence and absence of the inhibitors dibucaine or fluoride. Inhibition numbers are determined by comparing dibucaine or fluoride inhibited enzyme activity with total enzyme activity determined simultaneously. It is estimated that one patient in 1,500 is susceptible to succinylcholine anesthetic mishap. Owing to this low pretest probability, patients are rarely preoperatively screened for inherited succinylcholine or mivacurium anesthetic sensitivity. Samples collected during succinylcholine-induced apnea are inadequate for the measurement of BChE activity or to determine phenotype as its presence leads to anomalously low activity results and erroneous inhibitor numbers. It is critical to wait at least 24 h before obtaining a sample. If the patient has been treated with fresh frozen plasma or exogenous cholinesterase, a period of at least 6 weeks should be allowed before the sample is collected.

2.5 Questions and answers

1. What is the proposed physiological function of BChE?
   (a) Critical role in neurotransmission
   (b) Serve as a reservoir for essential amino acids
   (c) Activates the complement system
   (d) Bind compounds that inhibit AChE
   (e) Serve as a carrier protein for fatty acids and certain hormones
2. Which statement is false regarding the diagnostic use of BChE and/or AChE?
   (a) Used to monitor exposure to toxic organophosphate compounds
   (b) BChE activity is commonly used in the USA to diagnose liver disease
   (c) The presence of AChE in amniotic fluid can be used to diagnose fetal neural tube defects

    (d) Used to monitor exposure to toxic carbamate compounds

    (e) Dibucaine inhibition can identify individuals with BChE variants that are unable to efficiently hydrolyze succinylcholine

3. Which of the following statements is true?

    (a) Serum BChE activity is primarily used to monitor for chronic poisoning

    (b) RBC-AChE activity is primarily used to monitor for acute poisoning

    (c) BChE activity is a surrogate marker for *in vivo* neurotoxicity

    (d) Exposure to anticholinesterase compounds generally results in BChE activity declining more rapidly than AChE activity

    (e) Both AChE and BChE assays are commonly used in hospitals

4. What would you expect to occur in patients who are homozygous for atypical BChE after exposure to succinylcholine?

    (a) Prolonged apnea of more than 2 h; dibucaine number more than 70

    (b) Prolonged apnea of more than 2 h; dibucaine number less than 20

    (c) Prolonged apnea of more than 2 h; Dibucaine number less than 70

    (d) Normal response; dibucaine number less than 70

    (e) Slightly prolonged apnea; dibucaine number between 40 and 70

5. What is the primary clinical use for measuring cholinesterase activity?

    (a) Forensics

    (b) Pretesting of patients to ensure a normal response to succinylcholine or mivacurium

    (c) Assess fetal neural tube defects

    (d) Assess liver function

    (e) Identify exposures to dangerous pesticides or chemical warfare agents

6. Which of the following statements is true about carbamate poisonings?

    (a) Are less severe and of shorter duration than organophosphate poisonings

    (b) Result in the irreversible inhibition of the cholinesterases in a process called aging

    (c) Duration of symptoms is usually longer than 24 h

    (d) Sarin and soman are examples of carbamates

    (e) Are more severe and of shorter duration than organophosphate poisonings

7. Which of the following is NOT a common symptom of anticholinesterase poisoning?

    (a) Miosis

    (b) Urinary incontinence

    (c) Lacrimation

    (d) Defecation

    (e) Mydriasis

8. Which of the following statements is true regarding cholinesterase laboratory testing?

    (a) BChE has a narrower reference range than AChE

    (b) Serum BChE is often used to monitor acute poisoning

    (c) AChE is easier to measure and thus is more commonly used in hospitals

    (d) BChE is a surrogate marker for *in vivo* neurotoxicity

    (e) RBC-AChE is used to assess acute poisoning

9.   What is the most commonly used laboratory assay to measure cholinesterase activity?

    (a) Immunoassay

    (b) Delta pH method

    (c) Ellman method

    (d) Radiolabelled microassay

    (e) Jaffe method

References

[1] Proceedings of the Tenth International Meeting on Cholinesterases. Chem Biol Interact 2010,187,1–440.

[2] Taylor P, Radic Z. The cholinesterases: from genes to proteins. Annu Rev Pharmacol Toxicol 1994,34,281–320.

[3] Massoulie J, Perrier N, Noureddine H, Liang D, Bon S. Old and new questions about cholinesterases. Chem Biol Interact 2008,175,30–44.

[4] Chatonnet A, Lockridge O. Comparison of butyrylcholinesterase and acetylcholinesterase. Biochem J 1989,260,625–34.

[5] Soliday FK, Conley YP, Henker R. Pseudocholinesterase deficiency. a comprehensive review of genetic, acquired, and drug influences. AANA J 2010,78,313–20.

[6] Goodall R. Cholinesterase: phenotyping and genotyping. Ann Clin Biochem 2004,41,98–110.

[7] Clark RF. Insecticides: organic phosphorus compounds and carbamates. In: Flomenbaum NE, Goldfrank LR, Hoffman RS, Howland MA, Lewin NA, Nelson LS, eds. Goldfrank's Toxicologic Emergencies. 8th ed. New York, NY, USA, McGraw-Hill, 2006,1497–522.

[8] Tuorinsky SD, ed. Medical aspects of chemical warfare. Washington DC, USA, Borden Institute, 2008,1–773.

[9] Wilson BW, Arieta DE, Henderson JD. Monitoring cholinesterases to detect pesticide exposure. Chem Biol Interact 2005,157–158,253–6.

[10] McQueen M. Clinical and analytical considerations in the utilization of cholinesterase measurements. Clin Chem Acta 1995,237,91–105.

[11] Panteghini M, Bais R, van Solinge WW. Enzymes. In: Burtis CA, Ashwood ER, Bruns DE, eds. Tietz Textbook of Clinical Chemistry and Molecular Diagnostics. 4th ed. St Louis, MO, USA, Elsevier Saunders, 2006,597–643.

[12] Darvesh S, Hopkins DA, Geula C. Neurobiology of butyrylcholinesterase. Nat Rev Neurosci 2003,14,131–8.

[13] Wu A. Tietz Clinical Guide to Laboratory Tests. 4th ed. St Louis, MO, USA, Elsevier Saunders, 2006,1–1798.

[14] Ellman GL, Courtney KD, Andres V Jr, Feather-Stone RM. A new and rapid colorimetric determination of acetycholinesterase activity. Biochem Pharmacol 1961,7,88–95.

[15] Jacobs DS, DeMott WR, Oxley DK. Jacobs and DeMott laboratory test handbook. 5th ed. Hudson, OH, USA, Lexi-Comp Inc., 2001,1–1031.

3 Aldolase

Edmunds Reineks, Joe M. El-Khoury and Sihe Wang

3.1 Case studies

3.1.1 Patient A

An 80-year-old Caucasian female with a past medical history of myocardial infarction, hyperlipidemia, congestive heart failure, polymyalgia rheumatica (PMR), and giant cell arteritis (GCA) presented to her rheumatologist with complaints of headache, temporal tenderness, and tinnitus. A flare-up of her GCA was suspected and her maintenance low-dose steroid therapy was increased to address her symptoms. Her laboratory tests included the following: creatine kinase (CK) 51 U/L, aldolase (ALD) 4.1 U/L (normal <7.7 U/L), and C-reactive protein (CRP) 7 mg/L (normal <10 mg/L). With the change in therapy, her symptoms improved over the next several weeks, but she was hospitalized after she developed atrial fibrillation.

Her other medications included antihypertensives with diuretics, a statin, digoxin, and anticoagulants. To address her persistent hyperlipidemia, her statin dose was increased 4 months ago. One week after she was discharged, she presented with a complaint of compromised ability to rise from a chair and difficulty in walking. Examination showed mild to moderate proximal weakness of the lower extremities. She did not have shoulder or pelvic girdle pain, and no complaints of fever, chills, headaches, or visual disturbances. Laboratory tests at the post-discharge visit showed: CK 104 U/L, ALD 11.2 U/L, and CRP 10 mg/L.

3.1.1.1 Discussion

This patient presented with muscle weakness that could arise from various etiologies. Her underlying PMR/GCA may present or recur with muscle pain or fatigue, among various other symptoms [1, 2]. However, elevated muscle enzymes in patients with symptoms of PMR would indicate a diagnosis of an inflammatory myopathy [2]. Although this patient had been treated with a statin for many years, the relatively recent increased dosage merits consideration as a cause of the new symptoms and laboratory findings. Statin-induced myopathy is well described and is associated with increases in serum levels of muscle enzymes [3]. In addition, the increased steroid therapy could result in a myopathy with proximal muscle weakness. Although elevated (i.e., above the reference range) muscle enzymes are not typically associated with chronic steroid-induced myopathy, increased CK levels have been reported in some cases [4, 5].

The rising, although normal, CK coupled with the elevated ALD indicated muscle damage, although not definitively either statin- or steroid-induced. Urinary creatinine excretion has been reported to increase in cases of steroid-induced myopathy [6], but this analysis was not obtained in this case. The statin therapy was discontinued and the steroid therapy was tapered to the patient's previous maintenance dose. Symptomatic improvement related to leg weakness occurred within days, and follow-up laboratory tests 3 weeks later showed a return to baseline levels for CK and ALD. However, the elevated CRP persisted. Alternative approaches for management of the patient's hyperlipidemia were being considered.

3.1.2 Patient B

A 41-year-old African-American female with a mild history of well-controlled asthma presented to her primary care physician with complaints of fatigue, increasing difficulty climbing stairs, and some pain in her ankle joints and thighs. The symptoms began 1 month previous and seemed to be worsening. Also, for the past 4 days, she had had a fever of 38.5°C and some difficulty swallowing foods, even soft foods such as mashed potatoes. On examination, her proximal muscle strength in her legs was 3/5, but she showed no other muscle weaknesses. There were no skin changes. She was on no daily medications and had an inhaler, which she used less than four times per year.

Laboratory tests showed: CK 655 U/L, ALD 17 U/L, lactate dehydrogenase (LDH) 449 U/L, and CRP 12 mg/L. Thyroid tests were normal.

3.1.2.1 Discussion

The patient was referred to rheumatology. Screening for antinuclear antibodies was positive, testing for anti-Jo-1 antibody was weakly positive. The remaining autoantibody evaluation was negative. Follow-up tests consisted of magnetic resonance imaging (MRI), electromyography (EMG), and muscle biopsy. MRI showed increased T2 signal in the quadriceps and deltoid muscles, indicating inflammation and edema. EMG evaluation showed polyphasic motor unit potentials of short duration, and spontaneous fibrillations, interpreted as myopathic changes. The biopsy showed lymphocytic inflammation surrounding and invading scattered muscle fibers. Immunostaining demonstrated that these lymphocytes were primarily T cells. A diagnosis of polymyositis was rendered and steroid therapy was initiated [7, 8]. The lack of skin involvement or Raynaud's phenomenon reduced the likelihood of dermatomyositis, another diagnostic consideration [9, 10]. Inclusion bodies were not observed on the biopsy specimen.

Further tests included detailed evaluation of lung function and various screens for malignancy. There was currently no evidence of interstitial lung disease or malignancy.

The patient's symptoms were substantially reduced within 6 months of initiating therapy. At that time, muscle enzyme levels were reduced from those at presentation, although still elevated (CK 225 U/L, ALD 9.8 U/L, and LDH 250 U/L).

3.1.3 Patient C

A 52-year-old Caucasian male presented with generalized pain and swelling in both legs. The symptoms had persisted and worsened for approximately 3 months. He took no medications and was generally otherwise active and healthy. His surgical history consisted only of a left knee ligament repair 22 years ago. On examination, non-pitting edema was identified bilaterally on his legs. There was some skin tension leading to mildly restricted movement. When either leg was elevated from the examination table, visible longitudinal indentations were evident that corresponded to the courses of superficial veins. Focally, there were areas of skin thickening and puckering.

Laboratory tests were negative for thyroid and renal disorders. The complete blood count showed a white blood cell count elevated to 16,000/mm^3, with increased eosinophils to 29%. The erythrocyte sedimentation rate (ESR) was elevated at 66 mm/h. Evaluations for antinuclear antibodies and rheumatoid factors were negative. Serum immunoglobulin G levels were elevated (19,000 mg/L), electrophoresis showed the hypergammaglobulinemia to be polyclonal in nature. Other tests included: CK 77 U/L, ALD 10.2 U/L, and LDH 212 U/L.

3.1.3.1 Discussion
A full thickness skin biopsy, which included a small amount of superficial muscle tissue, was performed. It demonstrated inflammation in the subdermal and fascial layers, consisting of eosinophils, lymphocytes, and plasma cells. There was scant involvement in the epimysial layer. A diagnosis of eosinophilic fasciitis was rendered [11–13]. Steroid therapy was initiated. In follow-up tests over the next several months, the patient's white blood cell count and peripheral blood eosinophilia resolved. His ESR also normalized and most areas of the affected skin began to soften. Additional therapeutic measures, including psoralen UVA photochemotherapy, cyclosporine A, or intravenous immune globulin, have been reported to be of some success [13–15].

3.2 Biochemistry and physiology

3.2.1 Physiological function

ALD (EC 4.1.2.13), or specifically D-fructose-1,6-bisphosphate D-glyceraldehyde-3-phosphate-lyase, plays a central role in glycolysis, the breakdown of glucose to lactate. It acts by catalyzing the splitting of D-fructose-1,6-bisphosphate (FBP) into two triose phosphates, D-glyceraldehyde-3-phosphate (GLAP) and dihydroxyacetone-phosphate (DAP). In the reversible reaction, the enzyme is involved in gluconeogenesis.

3.2.2 Biochemistry and molecular structure

ALD is a tetramer with four equal subunits. Depending on the isoenzyme composition, ALD has a molecular weight of 150–160 kDa [16, 17]. ALD is cytosolic and has been shown to bind to multiple other cellular components, including F-actin and other structural proteins [18, 19]. The tetrameric form of ALD from various vertebrate sources is stable, with the subunits remaining associated even in dilute solutions [20]. Subunits can be separated by point mutations of interacting amino acids or under denaturing conditions [17, 20]. When separated under non-denaturing conditions, rabbit ALD activity remained in the individual subunits [17]. Each subunit has a structure that is described as consisting of an eight-stranded αβ-barrel with a centrally located active site [3, 21, 22].

Electrophoretic separation and measurement of activity, originally performed using ALD from rabbits, allowed distinction between the various isoforms of the enzyme [16]. Three distinct subunits (A, B, and C) have been identified, with different expression patterns based on tissue type in humans and other vertebrates [23]. The isoenzymes of ALD share considerable sequence homology, ranging in sequence identity from 66% to 78% between any pair of the three isoenzymes [19]. The amino acid residues at the active sites are conserved, but the activity of the isoenzymes can be distinguished based on their catalytic properties with FBP as the substrate or an alternate substrate, fructose-1-phosphate (F1P), which is structurally similar [16]. The kinetic analysis of the three isoenzymes showed increased activity by aldolase B on the F1P substrate, allowing it to be distinguished from the other isoenzymes [16]. Other differences were observed in V_{max} and K_m values. A radioimmunoassay, with the ability to distinguish the three isoenzymes, has also been developed [23].

As a central enzyme in glycolysis, the tissue expression of ALD is ubiquitous. A separate gene encodes for each isoenzyme [24]. Aldolase A, coded by the gene *ALDOA* on chromosome 16, is highly expressed in skeletal muscle and red blood cells [24]. Aldolase B is highly expressed in the liver and is also expressed in the small intestine and pancreas together with aldolase A [24]. The gene for aldolase B, *ALDOB*, is located on chromosome 9 [24]. Aldolase C is expressed in brain and other cells of the nervous system [25], its gene is *ALDOC* and is located on chromosome 17 [24].

The isoenzymes show diversity in their amino acid sequences in their C-terminal regions, which may account for differences in substrate selectivity [24]. The difference in activities between aldolase B, found in normal human liver, and aldolase A, typically found in muscle but also in cells of hepatocellular carcinoma (HCC), was utilized in a study of therapy for HCC [26]. The catalytic rate of aldolase A is 50-fold greater using FBP as a substrate versus F1P [16]. Aldolase B is equally active with both substrates [16]. In this study, investigators used ultrasound-guided, percutaneous ethanol injection to damage HCC cells found in liver nodules. To assess the damage to tumorous versus non-tumorous liver cells, serum ALD activity levels were assayed using FBP and F1P as substrates. Because damaged cells released some of their contents into the patient's bloodstream, damage to nearby normal liver tissue was assessed by serum ALD activity with F1P as the substrate [26]. Elevations in ALD activity with FBP as the substrate reflected damage to HCC cells and were associated with decreasing α-fetoprotein levels, a biomarker of HCC [26].

3.2.3 Tissue source(s) and expression of ALD

ALD, as a key component of glycolysis, is expressed ubiquitously [24]. However, as might be expected, its expression and activity are regulated depending on physiological status and tissue type. In one study using rabbit liver, investigators compared the enzyme activity of ALD in extracts from the livers of fed versus fasting rabbits, and they observed that ALD extracted from the livers of fed rabbits had at least twice the activity of that from fasting rabbit livers [27]. However, the concentration of ALD was the same in extracts of these tissues [27]. This finding showed that factors other than the amount of expressed ALD protein governed its activity, such as phosphorylation or some other post-translational modification. However, in liver tissues of a study in rats, the fed state was shown to induce expression level of aldolase B mRNA, as well as mRNA of several other glycolytic enzymes [28]. In a more comprehensive examination of rat aldolase B mRNA expression, investigators showed that carbohydrates can induce aldolase B mRNA in liver tissue of fed rats, provided they had a normal hormonal milieu [29]. By contrast, the hormone status of the animals did not impact the fructose-induced aldolase B mRNA in the renal tissues of the animals [29]. The hormonal dependence of aldolase B induction in small intestinal tissue was intermediate [29]. In a cell culture model, aldolase A transcriptional regulation was shown to be linked to the hypoxia inducible factor-1 (HIF-1) and was induced when hypoxic stressors were present or when HIF-1 was induced through other mechanisms [30]. In a study that sought to identify genes important in the development of malignant pleural effusions (MPEs), aberrant expression of aldolase A, as well as several other enzymes related to glucose metabolism, were identified [31].

3.2.4 Clearance and metabolism of ALD

The mechanism of ALD clearance from the circulation is not clearly established. However, given its size of 150–160 kDa, and its reluctance to dissociate from the tetrameric state, renal clearance is unlikely. For example, in a study that investigated the clearance of numerous enzymes released during muscle necrosis, large amounts of malate dehydrogenase and CK, at 62 kDa and 81 kDa, respectively, were excreted in the urine, whereas LDH, at 140 kDa, was not present in significant amounts in urine [32]. After infusion of plasma containing elevated levels of ALD and other enzymes, serum ALD levels peaked 2 days later [32]. By day 7, serum ALD was approximately

10% of its peak concentration, giving an approximate half-life estimate of 1–2 days [32]. In a more recent examination, it was found that rising serum ALD levels after a myocardial infarction peaked at 24–48 h, and returned to normal over the next 5 days [33, 34]. A similar estimate of the half-life may be inferred from these results.

Animal studies have also examined the clearance of ALD. An early study using a rat model suggested a short half-life if 12 h or less, as well as a role for renal excretion of ALD [35]. Using a mouse model system with the ability to block enzyme uptake by the reticuloendothelial system (RES), investigators found that the removal of ALD was unaffected by altering RES function [36]. This was in contrast to the effect on LDH plasma levels, which increased in response to RES inhibition [36]. In a study that utilized a perfused rat liver model system, uptake and degradation of ALD by the liver was demonstrated [37]. The rate of uptake into the liver tissue was dependent on the form of ALD, native ALD was removed slower than glutathione-inactivated ALD, but cathepsin D-inactivated ALD had the slowest uptake [37]. The rate of degradation after uptake was not dependent on the form of ALD [37]. Many questions regarding the elimination of ALD from the plasma, particularly in humans, remain to be addressed.

3.2.5 Reference ranges

Ranges typically quoted for ALD range from 1 to 8 U/L in the serum of adult patients. Owing to either increased muscle turnover or reduced clearance mechanisms, children have higher levels of ALD [38–40]. Healthy infants may have levels up to four times the adult concentration, and these will decrease through development [39].

ALD activity has been measured in the cerebrospinal fluid (CSF) as a possible indicator for damage/leakage from neuronal cells. Patients with acceleration-deceleration brain injuries were separated into two groups, with one group suffering mild brain damage (loss of consciousness <1 h) and the other group with severe brain injury (coma or semicomatose). Patients with more severe brain injuries were found to have higher levels of ALD in the CSF as well as serum [41]. ALD in the CSF was generally lower than corresponding serum levels [41]. The author concluded that CSF enzyme levels are not suitable as a prognostic tool. The small sample study size and lack of information about enzyme levels in various regions of the brain were cited as limitations [41]. Many additional tools have become available and established for clinical use since the time of this study; today, improved neuroimaging methods and established scales based on clinical evaluation and history of injured patients are the recognized methods for diagnosis and prognosis in brain injury patients [42].

Aldolase A expression, measured at the RNA level, appears to be altered in cells from patients with malignant pleural effusions [31]. However, the relationship of this overexpression to the actual ALD activity in the fluid is not clear. Thus, molecular diagnostic testing in this setting may contribute to the diagnosis/treatment of the patient, whereas the role of enzymology, if any, remains to be determined.

3.3 Chemical pathology

Patients presenting with complaints of fatigue or weakness have a broad differential diagnosis. Their disorder may be metabolic, psychological, or neurological. If the weakness is confirmed, two branches in the diagnostic tree include weakness of muscular origin versus neurological origin. The elevation of circulating muscle enzymes, such as ALD, CK, or LDH, provides a clue to the next diagnostic steps if the weaknesses are of muscular origin.

The value of measuring serum ALD activity as an indication of muscle damage was largely superseded by measurement of CK activity [43, 44]. Despite the emergence of cardiac troponins as the gold standard biomarker for diagnosis of myocardial damage, CK assays remain widely available due to many years when CK and CK-MB (the cardiac isoenzyme) combined to serve as the myocardial damage gold standard. Regarding muscle damage in general, some investigators indicate that CK assays have superior analytical performance as well as higher sensitivity and specificity for muscle injury [43, 45]. Indeed, given the expression of ALD in most other tissues, some of which are also susceptible to damage and leakage of enzymes, specificity is a potential area of concern. Nevertheless, ALD continues to be a useful marker of disease in spite of the perceived superiority of CK. Because the use of ALD activity is described in certain pathological settings in Section 3.1, areas where its utility is noteworthy will be emphasized.

3.3.1 Polymyositis/dermatomyositis

One of the earliest uses for ALD measurement was in the diagnosis and monitoring of inflammatory myopathies [40, 46]. Polymyositis and dermatomyositis are inflammatory conditions, likely to be of autoimmune etiology, which result in, among other symptoms, damage to muscle tissue [7, 47]. The diagnostic criteria are evolving and consist of a constellation of clinical signs/symptoms, as well as biochemical indications of muscle damage and inflammation [7, 47]. Therapy is often based on use of steroids, taking advantage of their anti-inflammatory activities. In addition to clinical improvement, resolution of biochemical abnormalities, such as lowering of CK, ALD, and LDH activities are signs of successful intervention [47].

This is one setting where continued use of ALD testing seems to be indicated. There are several reports of either individual cases or case series of patients with polymyositis or dermatomyositis where CK was normal, but ALD levels were elevated [9, 10, 48]. Some investigators estimate that up to 10% of patients with these disorders may not have elevated CK [9, 49]. Although the sensitivity of CK for diagnosis of inflammatory muscle disorders is high, adding ALD activity to the initial diagnostic workup may further increase the diagnostic sensitivity.

3.3.2 Duchenne muscular dystrophy

Another early use of ALD measurements was in the diagnosis and monitoring of muscular dystrophies [40, 46, 50]. Increased ALD is seen early in the disease, contributing to the diagnostic process and monitoring of disease progression [46]. However, in later stages of the disease, where muscle atrophy has significantly progressed, there is less leakage of muscle enzymes into the circulation because muscle mass is reduced. ALD and other muscle enzymes lose their usefulness at this point in following the pathological process [46].

3.3.3 Drug-induced myopathy

There are varied definitions of statin-induced myopathy, but this entity may consist of a spectrum of muscle disorders, including myalgia, myositis, or rhabdomyolysis [3, 51]. Guidelines from several organizations distinguish between myalgias and myositis on the basis of a rise in CK, which can be detected in myositis [51]. Some investigators have suggested the additional use of ALD in detecting or monitoring the development of statin-related myopathy [3, 52], particularly in cases where serum CK values may not correspond with the extent of muscle damage [3]. Given the large numbers of patients receiving statin therapy, evaluating and understanding the role of biochemical findings in these patients is essential to averting undesirable consequences of this common side effect [51]. Although initial studies regarding statin-induced myopathy suggested a low percentage of patients would be affected, subsequent data have shown that up to 9% of non-diabetic patients may experience a myopathic event while on statin therapy [3]. Additional affected patients may be identified if baseline testing were established at the initiation of therapy. This is based on the concern that the reference interval for CK and ALD may be too broad to appropriately identify patients who are experiencing statin-related muscle damage [3]. Reference change intervals, which measure the intra-individual variability of an analyte over time, were determined for CK and ALD in a healthy population who were not on statin therapy [3]. This work showed that serial changes in CK or ALD to 140% of baseline levels in patients on statin therapy should be considered significant and an indication of muscle damage [3]. The development of statin-induced myopathy may also arise with use of concomitant medications or substances, including verapamil, amiodarone, or large amounts of grapefruit juice [53].

Steroid-induced myopathy may also cause increases in circulating muscle enzymes [4, 54]. In this pathology, however, the acuteness of onset plays a role in the diagnostic utility of biochemical markers. In slowly developing chronic steroid myopathy, it is likely that no biochemical changes, including those in ALD, will be detected [4]. However, in individuals with more acute presentation, muscle enzymes can be elevated [4, 54]. A noted exception to this general guideline is reported from a series of five patients with juvenile dermatomyositis, who developed enzyme elevations after long-term steroid therapy [4].

Many other drug-induced myopathies have been reported. The drugs include various antimalarial agents and others [55, 56]. The diagnosis depends on the clinical presentation and, in many cases, results of a muscle biopsy demonstrating myotoxicity [55]. Elevated muscle enzymes may or may not be present, but should be included in a thorough evaluation to rule out other pathologies that may be contributing to the presentation [55, 56].

3.3.4 Pathologies in which CK does not reflect the extent of muscle damage

There may be patients in whom identifying and monitoring muscle damage is indicated, but CK is not a useful marker. One possible scenario is in the setting of a patient with macro-CK. In this condition, CK levels as assayed by routine clinical methods are elevated in patients without related pathological conditions, that is, no known muscle pathology [57]. Macro-enzymes of other types have been identified [e.g., macro-aspartate aminotransferase (macro-AST)] where the source tissue(s) (liver) are not damaged according to other diagnostic evaluations [57]. Certain macro-enzymes are found to be bound to anti-enzyme immunoglobulins and, although unlikely, their presence may indicate that the patient is at risk of developing some condition, such as an autoimmune disorder [57]. In a patient with macro-CK, ALD and LDH may provide alternative means to identify and follow developing muscle pathologies.

Another study identified patients in a critical care setting where serum CK activity levels were unexpectedly low, despite clinical and other biochemical evidence of muscle wasting [58]. Many critically ill patients, especially those with liver disease, may have low levels of glutathione, an endogenous antioxidant. CK has multiple thiol groups that are susceptible to oxidation, and optimal enzyme activity is present only when these groups are in the reduced form [58]. A correlation between low levels of glutathione and low CK activity, in the presence of elevated ALD and myoglobin, demonstrated the unreliability of using CK to assess muscle damage or wasting in this setting [58].

A case series consisting of 12 patients was published to demonstrate settings in which patients were experiencing muscle discomfort or weakness without an increase in serum CK but with elevated serum ALD [59]. The pathological processes involved covered a broad range of conditions, including respiratory distress syndrome, eosinophilia, vasculitis, rheumatoid arthritis, graft versus host disease, and dermatomyositis [59]. Diagnostic and/or follow-up studies showed that perimysial inflammation was a common finding in these patients [59]. The authors suggest measuring serum ALD in the setting of an otherwise negative neuromuscular evaluation to help identify patients that might benefit from a muscle biopsy [59]. An earlier retrospective study examined the results and diagnoses of 126 consecutive patients who had serum ALD measurements, most of whom also had CK performed on these same specimens [48]. Of these patients, 19 had increased ALD but normal CK [48]. Their diagnoses included rheumatoid arthritis, PMR, polymyositis, systemic lupus erythematous, fibromyalgia, and various other conditions [48].

In summary, despite the increased specificity and availability of CK as an indicator of muscle damage, determination of serum ALD has been shown to add to the diagnostic sensitivity of various conditions. ALD should remain a consideration when evaluating patients presenting with fatigue or weakness.

3.4 Analysis

Two methods were originally described to measure ALD activity. A colorimetric method utilized hydrazine to trap the cleavage products of fructose 1,6-bisphosphate, forming complexes with GLAP and DAP [60, 61]. After quenching the ALD reaction with acid, these complexes could be decomposed with the use of alkaline reagents, and a colorimetric complex formed by reaction with 2,4-dinitrophenylhydrazine. Measurement of absorption at 540 nm was converted to enzyme activity level [60, 61].

The second method coupled the oxidation of NADH to the reduction of DAP, one of the triosephosphate cleavage products of ALD [62, 63]. Initially, the triosephosphate products consist of a 1:1 mixture of GLAP and DAP. In this assay, triosephosphate isomerase (EC 5.3.1.1), which strongly favors the formation of DAP over GLAP, was added to drive this conversion. Then, glycerol-3-phosphate dehydrogenase (EC 1.1.1.8) was used to reduce DAP with the oxidation of NADH. Thus, 2 mol of NADH was oxidized per cleaved mole of fructose 1,6-bisphosphate [62, 63].

Other methods to monitor ALD activity have been described. For example, there is a method that couples the triosephosphate product formation to reduction of cytochrome *c*, which can be monitored spectrophotometrically. However, the method makes use of a cyanide reagent, and there is no evidence that this method is currently in clinical use [64].

Commercial kits for diagnostic use are available, and these measure ALD activity by the second (coupled) UV-monitored reaction [65, 66].

3.4.1 Specimen

One commercially available kit indicates that the assay should be performed on serum, but other kits allow use of serum or plasma [65, 66]. Some reference laboratories will accept only serum specimens, whereas others will accept serum or plasma, containing citrate, oxalate, EDTA, or heparin [67–69]. It has been suggested that plasma activity may better reflect *in vivo* activity due to release of high concentrations of ALD from platelets during clotting [70, 71]. By contrast, a more recent comparison of serum and heparinized-plasma ALD was performed using an automated analyzer, with markedly different results [72]. Average levels of plasma ALD were 39% higher than serum samples collected simultaneously from 20 healthy subjects [72]. LDH showed a similar pattern, suggesting that a common source of both enzymes, erythrocytes or platelets, were suggested to have contaminated the plasma samples to a higher degree than the serum samples. Although this explanation was proposed by the investigators, it was not proven [72]. Explanations for the observed differences between serum and plasma appear to focus on pre-analytical differences (the clotting process, RBC or platelet contamination, etc.) rather than on differences during the analytical phase [70, 72]. As with many analytes, laboratories are wise to establish their own reference ranges to account for differences in specimen collection and handling, specimen type, patient population, etc.

3.4.2 Analyte stability

Whether plasma or serum specimens are assayed, prompt separation of the serum or plasma from the cells and/or clot needs to be emphasized. After separation, the analyte is stable for at least 8 h at room temperature, for 5 days if refrigerated, and up to 6 months when frozen [67, 68].

3.4.3 Interferences

Hemolysis is clearly an indication for rejection in cases where muscle damage is being evaluated [70]. This is because erythrocytes, which are glycolytically active, contain large amounts of ALD. In such samples, the measured ALD activity does not correspond to the amount of muscle damage.

3.4.4 Reference methods

A reference method has not been identified. Despite the establishment of reference methods and materials for several other clinically significant enzymes, the waning role of ALD in clinical settings may have contributed to its exclusion from the standardization process [73]. The second method described above, linking DAP reduction to NADH oxidation, is established as the primary clinical method. This method was evaluated and improved in a process that addressed some of its early limitations [62].

3.5 Questions and answers

1. Serum aldolase is typically highest in which of the following groups?
   (a) Healthy infants
   (b) Adult males
   (c) Adult females
   (d) Post-menopausal females
   (e) Geriatric patients
2. Which of the following enzymes has supplanted aldolase as a diagnostic tool?
   (a) Lactate dehydrogenase
   (b) Aspartate aminotransferase
   (c) Alanine aminotransferase
   (d) Creatine kinase
   (e) Alkaline phosphatase
3. Which of the following is NOT true about aldolase?
   (a) It is found in the cytosol
   (b) Three distinct isoenzymes (A, B and C) have been identified
   (c) Its subunits consist of an eight stranded αβ-barrel with a centrally located active site
   (d) It is a tetramer consisting of four distinct subunits
   (e) It can bind to F-actin and other structural proteins inside the cell
4. The isoenzymes of aldolase differ from each other in all of the following EXCEPT?
   (a) Tissue expression
   (b) Amino acid residues at the active site
   (c) V_{max} and K_m values
   (d) Amino acid sequence of the C-terminal region
   (e) Catalytic activity
5. Elevated levels of aldolase generally indicate what?
   (a) Renal failure
   (b) Pancreatitis
   (c) Muscle damage
   (d) Diabetes
   (e) All of the above
6. What enzymes are elevated in muscular damage?
   (a) CK, ALD, and LDH
   (b) CK, LDH and AST
   (c) ALD, LDH and AST
   (d) CK, ALD, and AST
   (e) CK, ALD, LDH and AST

7. Measurement of serum aldolase is helpful in all of the following conditions EXCEPT?
   (a) Muscular dystrophy
   (b) Statin-induced myopathy
   (c) Pancreatitis
   (d) Polymyositis
   (e) Dermatomyositis
8. Which of the following specimens is unacceptable for the measurement of serum aldolase activity?
   (a) Serum separator
   (b) EDTA plasma
   (c) Plain serum
   (d) Sodium heparin
   (e) None
9. What methods have been developed for the measurement of serum aldolase activity?
   (a) Turbidimetric
   (b) pH-stat
   (c) Spectrophotometric
   (d) Immunochemical
   (e) High-performance liquid chromatography
10. Which of the following significantly increases serum aldolase?
    (a) Hemolysis
    (b) Lipemia
    (c) Icterus
    (d) Uremia
    (e) None of the above

References

[1] Salvarani C, Cantini F, Hunder GG. Polymyalgia rheumatica and giant-cell arteritis. Lancet 2008,372,234–45.
[2] Brooks RC, McGee SR. Diagnostic dilemmas in polymyalgia rheumatica. Arch Intern Med 1997,157,162–8.
[3] Wu AH, Smith A, Wians F. Interpretation of creatine kinase and aldolase for statin-induced myopathy: reliance on serial testing based on biological variation. Clin Chim Acta 2009,399,109–11.
[4] Naim MY, Reed AM. Enzyme elevation in patients with juvenile dermatomyositis and steroid myopathy. J Rheumatol 2006,33,1392–4.
[5] Hanson P, Dive A, Brucher JM, Bisteau M, Dangoisse M, Deltombe T. Acute corticosteroid myopathy in intensive care patients. Muscle Nerve 1997,20,1371–80.
[6] Askari A, Vignos PJ Jr, Moskowitz RW. Steroid myopathy in connective tissue disease. Am J Med 1976,61,485–92.

[7] Dalakas MC, Hohlfeld R. Polymyositis and dermatomyositis. Lancet 2003,362,971–82.

[8] Bohan A, Peter JB, Bowman RL, Pearson CM. Computer-assisted analysis of 153 patients with polymyositis and dermatomyositis. Medicine (Baltimore) 1977,56,255–86.

[9] Carter JD, Kanik KS, Vasey FB, Valeriano-Marcet J. Dermatomyositis with normal creatine kinase and elevated aldolase levels. J Rheumatol 2001,28,2366–7.

[10] Mercado U. Dermatomyositis with normal creatine kinase and elevated aldolase levels. J Rheumatol 2002,29,2242.

[11] Quintero-Del-Rio AI, Punaro M, Pascual V. Faces of eosinophilic fasciitis in childhood. J Clin Rheumatol 2002,8,99–103.

[12] Nakajima H, Fujiwara S, Shinoda K, Ohsawa N. Magnetic resonance imaging and serum aldolase concentration in eosinophilic fasciitis. Intern Med 1997,36,654–6.

[13] Horacek E, Sator PG, Gschnait F. 'Venous furrowing': a clue to the diagnosis of eosinophilic fasciitis. A case of eosinophilic fasciitis ultimately treated with oral PUVA therapy. Dermatology 2007,215,89–90.

[14] Bukiej A, Dropinski J, Dyduch G, Szczeklik A. Eosinophilic fasciitis successfully treated with cyclosporine. Clin Rheumatol 2005,24,634–6.

[15] Pimenta S, Bernardes M, Bernardo A, Brito I, Castro L, Simoes-Ventura F. Intravenous immune globulins to treat eosinophilic fasciitis: a case report. Joint Bone Spine 2009,76,572–4.

[16] Rutter WJ, Rajkumar T, Penhoet E, Kochman M, Valentine R. Aldolase variants: structure and physiological significance. Ann NY Acad Sci 1968,151,102–17.

[17] Beernink PT, Tolan DR. Disruption of the aldolase A tetramer into catalytically active monomers. Proc Natl Acad Sci USA 1996,93,5374–9.

[18] Arnold H, Pette D. Binding of aldolase and triosephosphate dehydrogenase to F-actin and modification of catalytic properties of aldolase. Eur J Biochem 1970,15,360–6.

[19] Arakaki TL, Pezza JA, Cronin MA, Hopkins CE, Zimmer DB, Tolan DR, et al. Structure of human brain fructose 1,6-(bis)phosphate aldolase, linking isozyme structure with function. Protein Sci 2004,13,3077–84.

[20] Lebherz HG. Stability of quaternary structure of mammalian and avian fructose diphosphate aldolases. Biochemistry 1972,11,2243–50.

[21] Dalby A, Dauter Z, Littlechild JA. Crystal structure of human muscle aldolase complexed with fructose 1,6-bisphosphate: mechanistic implications. Protein Sci 1999,8,291–7.

[22] Dalby AR, Tolan DR, Littlechild JA. The structure of human liver fructose-1,6-bisphosphate aldolase. Acta Crystallogr D Biol Crystallogr 2001,57,1526–33.

[23] Asaka M, Alpert E. Subunit-specific radioimmunoassay for aldolase A, B, and C subunits: clinical significance. Ann NY Acad Sci 1983,417,359–67.

[24] Tolan DR, Niclas J, Bruce BD, Lebo RV. Evolutionary implications of the human aldolase-A, -B, -C, and -pseudogene chromosome locations. Am J Hum Genet 1987,41,907–24.

[25] Inagaki H, Haimoto H, Hosoda S, Kato K. Aldolase C is localized in neuroendocrine cells. Experientia 1988,44,749–51.

[26] Khan KN, Nakata K, Nakao K, Kato Y, Eguchi K. Use of FDP and F1P aldolase to detect tumorous and nontumorous tissue damage by ethanol injection of hepatocellular carcinoma. Dig Dis Sci 1999,44,1610–8.

[27] Pontremoli S, Melloni E, Salamino F, Sparatore B, Michetti M, Horecker BL. Changes in activity of fructose-1,6-bisphosphate aldolase in livers of fasted rabbits and accumulation of crossreacting immune material. Proc Natl Acad Sci USA 1979,76,6323–5.

[28] Koo HY, Wallig MA, Chung BH, Nara TY, Cho BH, Nakamura MT. Dietary fructose induces a wide range of genes with distinct shift in carbohydrate and lipid metabolism in fed and fasted rat liver. Biochim Biophys Acta 2008,1782,341–8.

[29] Munnich A, Besmond C, Darquy S, Reach G, Vaulont S, Dreyfus JC, et al. Dietary and hormonal regulation of aldolase B gene expression. J Clin Invest 1985,75,1045–52.

[30] Semenza GL, Roth PH, Fang HM, Wang GL. Transcriptional regulation of genes encoding glycolytic enzymes by hypoxia-inducible factor 1. J Biol Chem 1994,269,23757–63.
[31] Lin CC, Chen LC, Tseng VS, Yan JJ, Lai WW, Su WP, et al. Malignant pleural effusion cells show aberrant glucose metabolism gene expression. Eur Respir J 2010,9,30.
[32] Dawson DM, Alper CA, Seidman J, Mendelsohn J. Measurement of serum enzyme turnover rates. Ann Intern Med 1969,70,799–805.
[33] Taguchi K, Takagi Y. [Aldolase]. Rinsho Byori 2001,Suppl 116,117–24 (in Japanese).
[34] Brancaccio P, Lippi G, Maffulli N. Biochemical markers of muscular damage. Clin Chem Lab Med 2010,48,757–67.
[35] Sibley JA. Significance of serum aldolase levels. Ann NY Acad Sci 1958,75,339–48.
[36] Mahy BW, Rowson KE, Parr CW, Salaman MH. Studies on the mechanism of action of Riley virus. I. Action of substances affecting the reticuloendothelial system on plasma enzyme levels in mice. J Exp Med 1965,122,967–81.
[37] Bond JS, Aronson NN Jr. Endocytosis and degradation of native, cathepsin D-degraded, and glutathione-inactivated aldolase by perfused rat liver. Arch Biochem Biophys 1983,227,367–72.
[38] Mayo Medical Laboratories. Accessed 6 December 2010. Available from: http://www.mayomedicallaboratories.com/test-catalog/print.php?unit_code#equal#8363.
[39] Visnapuu LA, Karlson LK, Dubinsky EH, Szer IS, Hirsch CA. Pediatric reference ranges for serum aldolase. Am J Clin Pathol 1989,91,476–7.
[40] Clayton BE, Wilson KM, Carter CO. Aldolase activity in the plasma or serum of normal children and families with muscular dystrophy. Arch Dis Child 1963,38,208–14.
[41] Klun B. Spinal fluid and blood serum enzyme activity in brain injuries. J Neurosurg 1974,41,224–8.
[42] Butcher I, McHugh GS, Lu J, Steyerberg EW, Hernandez AV, Mushkudiani N, et al. Prognostic value of cause of injury in traumatic brain injury: results from the IMPACT study. J Neurotrauma 2007,24,281–6.
[43] Hood D, Van Lente F, Estes M. Serum enzyme alterations in chronic muscle disease. A biopsy-based diagnostic assessment. Am J Clin Pathol 1991,95,402–7.
[44] Mandell BF. Aldolase in the diagnosis of myositis. Am J Med 1991,90,662.
[45] Wu AH, Perryman MB. Clinical applications of muscle enzymes and proteins. Curr Opin Rheumatol 1992,4,815–20.
[46] Hughes BP. Serum enzyme changes in muscle disease and their relation to tissue change. Proc R Soc Med 1963,56,179–82.
[47] Rider LG, Miller FW. Laboratory evaluation of the inflammatory myopathies. Clin Diagn Lab Immunol 1995,2,1–9.
[48] Tormey WP. Serum aldolase with creatine kinase in current clinical practice. Br J Clin Pract 1990,44,582–4.
[49] Liozon E, Vidal E, Sparsa A. Aldolase levels in dermatomyositis and polymyositis with normal creatine kinase levels. J Rheumatol 2003,30,2077–8.
[50] Mink JK, Greebe HM. Fructo-aldolase in relation to muscular dystrophies. Acta Neurol Scand 1964,40,107–14.
[51] Joy TR, Hegele RA. Narrative review, statin-related myopathy. Ann Intern Med 2009,150,858–68.
[52] Hyman MH. Issues in statin-associated myopathy. J Am Med Assoc 2003,290,888.
[53] Christopher-Stine L. Statin myopathy: an update. Curr Opin Rheumatol 2006,18,647–53.
[54] Khan MA, Larson E. Acute myopathy secondary to oral steroid therapy in a 49-year-old man: a case report. J Med Case Rep 2011,5,82.
[55] Valiyil R, Christopher-Stine L. Drug-related myopathies of which the clinician should be aware. Curr Rheumatol Rep 2010,12,213–20.
[56] Kalajian AH, Callen JP. Myopathy induced by antimalarial agents, the relevance of screening muscle enzyme levels. Arch Dermatol 2009,145,597–600.

[57] Remaley AT, Wilding P. Macroenzymes: biochemical characterization, clinical significance, and laboratory detection. Clin Chem 1989,35,2261–70.

[58] Gunst JJ, Langlois MR, Delanghe JR, De Buyzere ML, Leroux-Roels GG. Serum creatine kinase activity is not a reliable marker for muscle damage in conditions associated with low extracellular glutathione concentration. Clin Chem 1998,44,939–43.

[59] Nozaki K, Pestronk A. High aldolase with normal creatine kinase in serum predicts a myopathy with perimysial pathology. J Neurol Neurosurg Psychiatry 2009,80,904–8.

[60] Sibley JA, Lehninger AL. Determination of aldolase in animal tissues. J Biol Chem 1949,177, 859–72.

[61] Pinto PV, Van Dreal PA, Kaplan A. Aldolase. I. Colorimetric determination. Clin Chem 1969,15,339–48.

[62] Pinto PV, Kaplan A, Van Dreal PA. Aldolase. II. Spectrophotometric determination using an ultraviolet procedure. Clin Chem 1969,15,349–60.

[63] Racker E. Spectrophotometric measurement of hexokinase and phosphohexokinase activity. J Biol Chem 1947,167,843–54.

[64] Robertson P Jr, Fridovich I. Continuous colorimetric monitoring of the fructose bisphosphate aldolase reaction. Anal Biochem 1980,108,332–4.

[65] CaldonBiotech. Aldolase product insert. Accessed 6 December 2010. Available from: http://www.caldonbiotech.com/index2.html.

[66] Randox Laboratories Ltd. Aldolase reagent brochure, 2009. Accessed 5 May 2011. Available from: http://www.randox.com/brochures/PDF%20Brochure/LT194.pdf.

[67] Labcorp. Laboratory Corporation of America. Accessed 6 December 2010. Available from: http://www.labcorp.com/wps/portal/.

[68] ARUP Laboratories. Accessed 7 December 2010. Available from: http://www.aruplab.com/guides/ug/tests/0020012.jsp.

[69] MayoMedicalLaboratories. Aldolase, serum, 2011. Accessed 5 May 2011. Available from: http://www.mayomedicallaboratories.com/test-catalog/print.php?unit_code#equal#8363.

[70] Tietz NW, Burtis CA, Ashwood ER, Bruns DE. Tietz textbook of clinical chemistry and molecular diagnostics. 4th ed. St Louis, MO, USA, Elsevier Saunders, 2006.

[71] Dale RA. Demonstration of aldolase in human platelets. The relation to plasma and serum aldolase. Clin Chim Acta 1960,5,652–63.

[72] Miles RR, Roberts RF, Putnam AR, Roberts WL. Comparison of serum and heparinized plasma samples for measurement of chemistry analytes. Clin Chem 2004,50,1704–6.

[73] Panteghini M. Standardization in clinical enzymology, 2009. Accessed 5 May 2011. Available from: http://www.ifcc.org/.

4 Alkaline phosphatase

Amy E. Schmidt

4.1 Case studies

4.1.1 Patient A

A 35-year-old male presents to his primary care physician with complaints of fatigue, pruritus, abdominal pain, fevers, weight loss, and intermittent jaundice. Over the past 6 months he has lost 10 kg and his pruritus has been getting increasingly worse. He denies shortness of breath and his appetite has not changed. He has had some diarrhea that is occasionally associated with blood. He has not had any changes in his vision or hearing. He has never smoked and only drinks 2–3 beers on the weekends. He has never used illicit drugs and is married with one child. His father, paternal grandmother, and sister all have ulcerative colitis.

Physical examination shows a thin male with scleral icterus and pale mucus membranes. His vital signs are normal, but he has a systolic ejection murmur. He has hepatomegaly with the edge of the liver being 5 cm below the costal margin. Upon examining his skin, excoriations are seen on his arms, legs, chest, and back.

Laboratory tests performed on blood collected at his office visit show: hemoglobin 91 g/L, hematocrit 27%, white blood cell count 9.8×10^9/L, platelets 219×10^9/L, aspartate aminotransferase (AST) 24 U/L, alanine aminotransferase (ALT) 27 U/L, alkaline phosphatase (ALP) 853 U/L, 5′-nucleotidase 63 U/L, γ-glutamyltransferase (GGT) 99 U/L, total bilirubin 73.5 μmol/L, direct bilirubin 63.3 μmol/L, prothrombin time (PT) 14.3 s/international normalized ratio (INR) 1.1, perinuclear anti-neutrophil cytoplasmic antibody (p-ANCA) positive, anti-*Saccharomyces cerevisiae* antibody positive.

4.1.1.1 Discussion

This patient showed signs of liver disease as well as inflammatory bowel disease. He had pruritus with excoriations, scleral icterus, pallor, intermittent jaundice, weight loss, and hepatomegaly all suggesting liver disease. The positive p-ANCA and anti-*S. cerevisiae* antibodies suggest this, whereas the weight loss and occasionally bloody diarrhea indicate that the patient has ulcerative colitis. His family history also supports this.

His normal ALT and AST values with elevated GGT, 5′-nucleotidase, ALP, bilirubin, and GGT suggest obstructive hepatobiliary disease. Although the ALP could have been from bone, the GGT and 5′-nucleotidase support a hepatic origin. Primary sclerosing cholangitis (PSC) is commonly seen in patients with ulcerative colitis [1, 2]. PSC is a cholestatic liver disease in which patients have extrahepatic strictures of their biliary tree. The etiology of PSC is unknown; it is progressive and usually leads to cirrhosis, portal hypertension, and liver failure [3]. When it causes end-stage liver disease, the only treatment is liver transplantation. Endoscopic retrograde cholangiopancreatography (ERCP) is the gold standard for diagnosing PSC and can also be used to place a biliary stent, which can temporarily improve the PSC symptoms. The normal ALT and AST values indicate that no hepatic inflammation or cell death has yet occurred.

4.1.2 Patient B

A 62-year-old male presents with complaints of recurrent headaches, problems with hearing, and bone pain. He is upset that his motorcycle and bicycle helmets no longer fit. He denies fevers, chills, and night sweats. His weight and appetite are stable. He has no shortness of breath or heart palpations. His bone pain is particularly worse in his back, and his headaches occur at all times of the day.

Physical examination shows a well-developed, well-nourished white male with normal vital signs. His head appears somewhat square with some frontal bossing. His vision is normal. His hearing is impaired, as the doctor has to repeat a lot and speak very loud. His neck is supple with no lymphadenopathy, his spine shows kyphosis, his lungs are clear, and his heart is normal. There is no hepatosplenomegaly.

Laboratory tests performed on blood collected at his office visit show: hemoglobin 142 g/L, hematocrit 43.3%, white blood cell count 8.8×10^9/L, platelets 254×10^9/L, AST 24 U/L, ALT 27 U/L, ALP 2421 U/L, 5'-nucleotidase 9 U/L, GGT 15 U/L, total bilirubin 15.4 µmol/L, direct bilirubin 5.1 µmol/L, PT 14.0 s/INR 1.1, creatinine 70.7 µmol/L, serum C-telopeptide 1534 pg/mL (reference range = 87–345 ng/mL), and urine N-telopeptide 1,132 nM bone collagen equivalents/mmol creatinine (reference range = 5.4–24.3 nM bone collagen equivalents/mmol creatinine), serum calcium 2.25 mmol/L, phosphorous 1.03 mmol/L, and uric acid 625 µmol/L.

4.1.2.1 Discussion

This patient has bone pain, kyphosis (curving of the upper back), recurrent headaches, loss of hearing, and enlarged head with frontal bossing. This could indicate growth hormone excess, a pituitary or brain tumor, osteoarthritis, osteoporosis, or bone disease. The laboratory results show an elevated ALP with normal ALT, AST, GGT, and 5'-nucleotidase indicating that the elevated ALP is not from the liver. The elevated serum C-telopeptide and urine N-telopeptide support a high rate of bone turnover indicating that the ALP is probably bone-specific ALP. The serum calcium, phosphorous, and creatinine are all normal indicating that the problem is probably not due to hyperparathyroidism.

The patient was sent for X-ray to diagnose Paget's disease. Osteoporosis circumscripta was seen in the frontal and occipital bones of the skull; moreover, the "cotton wool" pattern characteristic of the mixed phase of Paget's disease was also seen in the skull. The patient's impaired hearing was attributed to bone compression of the eighth cranial nerve. Owing to the hearing loss and bone pain, the patient was started on bisphosphonates and will return to the clinic in 2 months for retesting.

Paget's disease is a disorder of bone remodeling. It begins with excess osteoclast activity or resorption followed by osteoblast activity laying down disordered, mosaic bone that is larger and less compact [4, 5]. The remodeled larger bones are weaker and more subject to fracture. Paget's disease usually affects older individuals and is much more common in Caucasians and individuals of European descent. Patients most frequently present with complaints of bone pain or simply with a markedly elevated ALP on routine laboratory testing.

4.1.3 Patient C

A 23-year-old G1P0 female at 35 weeks' gestation presents to her obstetrics and gynecology office for a routine appointment. She tells the doctor that she has noticed an increase in the swelling of her feet and legs and has felt dizzy several times after standing up quickly. She has no prior medical problems and has had no complications with the pregnancy to date.

Physical examination shows a well-appearing pregnant female. Her vital signs including her blood pressure (BP) are normal. She is checked for orthostatic hypotension and the BP values are normal. Mucus membranes are moist, no scleral icterus, the heart shows regular rate and rhythm with no murmur, and her lungs are clear. She has 1+ pitting edema on her bilateral feet and ankles. The baby's heart rate is normal and an ultrasound shows that everything also appears normal.

Laboratory tests performed on blood collected at her office visit show: hemoglobin 126 g/L, hematocrit 38.7%, white blood cell count 6.3×10^9/L, platelets 194×10^9/L, AST 31 U/L, ALT 29 U/L, ALP 163 U/L, total bilirubin 13.7 μmol/L, direct bilirubin 5.1 μmol/L, creatinine 61.0 μmol/L, serum calcium 2.25 mmol/L, phosphorous 1.0 mmol/L, sodium (Na) 140 mmol/L, potassium (K) 4.0 mmol/L, chloride (Cl) 101 mmol/L, bicarbonate 32 mmol/L and glucose 5.5 mmol/L.

4.1.3.1 Discussion
This patient is in the third trimester of her first pregnancy. Her physical examination is essentially normal and the ultrasound is also normal. The only abnormal laboratory result is ALP, which is elevated. During the third trimester of pregnancy, the placenta makes ALP, which elevates the mother's ALP level [6]. ALP is made in the liver, kidneys, bone, intestine, and placenta. The majority of ALP in the blood of a healthy adult is from the liver and bone; however, during pregnancy, placental ALP increases especially during the last trimester [7]. In rare cases, placental ALP is expressed by

some tumors. Importantly, the elevated ALP in the above patient is a normal aspect of pregnancy and nothing to be worried or concerned about. Moreover, it requires no further testing or evaluation.

4.1.4 Patient D

A 56-year-old female with a history of polycystic ovarian syndrome (PCOS), hypertension, and type II diabetes presents to the emergency department with polyuria, polydipsia, and dizziness. She is obese with a body mass index (BMI) of 38 kg/m^2. Her BP is 160/94 with a heart rate of 96 beats per minute. Her oxygen saturation is 99% on room air, and her respiratory rate is 17/min. On physical examination, she has several areas of apparent candida growth between rolls of skin as well as areas of acanthosis nigricans on her neck and arms. Her lungs are clear and her heart rate is tachycardic with no murmurs. Her abdomen is soft and non-tender. She has decreased sensation in her feet, with 1+ edema.

Laboratory tests reveal a hemoglobin of 82 g/L, hematocrit 25.3%, white blood cell count 11.1×10^9/L and platelets 221×10^9/L. Her other laboratory results also have numerous abnormalities: Na 144 mmol/L, K 5.5 mmol/L, Cl 103 mmol/L, bicarbonate 20 mmol/L, AST 15 U/L, ALT 20 U/L, ALP 194 U/L, total bilirubin 15.4 µmol/L, creatinine 137.3 µmol/L, blood urea nitrogen (BUN) 21.8 mmol/L, and glucose 28.9 mmol/L. Her blood type is B+.

4.1.4.1 Discussion
This patient is an obese female with PCOS, diabetes, and hypertension who presented to the emergency department with diabetic ketoacidosis. Her blood glucose was very elevated at 28.9 mmol/L. Aside from having an anion gap, elevated potassium, elevated WBC, and low hemoglobin and hematocrit, the patient has an elevated ALP in the setting of normal liver enzymes and normal bilirubin. This is probably attributable to several factors including the patient's obesity, increased glucose, and type B blood. With increasing weight, ALP, AST, and ALT values have been seen to increase in patients [8]. These values have been attributed both to obesity as well as to non-alcoholic fatty liver disease. Elevated glucose levels, diabetes, and type B blood all cause intestinal ALP (IAP) to be increased [9]. Increased IAP activity is also seen in chronic renal failure, liver cirrhosis, and diabetes. Moreover, patients with blood groups O also have increased IAP concentrations, which is especially true postprandially [10].

4.1.5 Patient E

An 18-year-old male college freshman presents to the emergency department with complaints of weight loss, fatigue, night sweats, fevers, and chills. This started

2 weeks ago, and he has lost 18 kg. He describes waking up several times each night covered in sweat necessitating changing the sheets at least once per night. His fevers have been as high as 39.7°C. He has never been hospitalized and has no significant past medical history. His mother had breast cancer, his father has coronary artery disease and myocardial ischemia as well as hypertension, his maternal grandfather died of a stroke, his maternal grandmother has glaucoma, his paternal grandfather died in a car accident, and his paternal grandmother died of acute myeloid leukemia.

His weight is 53 kg and he is 1.7 m tall, his BP is 120/76, his heart rate is 105/min, and respirations are 20/min. His oxygen saturation is 99% on room air. Temperature is 38°C. His mucus membranes are moist with no scleral icterus. His neck is supple, but several lumps are noted. He is tachycardic with a flow murmur, his lungs are clear, his abdomen is soft, non-tender and non-distended, with splenomegaly.

Laboratory tests reveal a hemoglobin of 85 g/L, hematocrit 26.1%, WBC 23.1×10^9/L (differential: 48% neutrophils, 24% lymphocytes, 18% monocytes, 8% eosinophils, and 2% basophils), and platelets 110×10^9/L. Other laboratory results also show numerous abnormalities: Na 144 mmol/L, K 4.5 mmol/L, Cl 108 mmol/L, bicarbonate 25 mmol/L, AST 52 U/L, ALT 61 U/L, ALP 352 U/L, total bilirubin 32.5 µmol/L, creatinine 83.9 µmol/L, BUN 7.5 mmol/L, lactate dehydrogenase (LDH) 280 U/L, and glucose 5.7 mmol/L.

Lymphoma is suspected and a lymph node is biopsied the next day. The lymph node shows a nodular sclerosing pattern with numerous Reed-Sternberg cells. Immunophenotyping showed that the involved cells were CD15, CD20, and CD30 positive.

4.1.5.1 Discussion

This patient is a male presenting with weight loss, fevers, chills, and night sweats, which are B symptoms of Hodgkin's lymphoma. Moreover, physical examination shows enlarged lymph nodes palpable in his neck, one of which was biopsied and shown to be Hodgkin's lymphoma. Often, patients with Hodgkin's lymphoma present without B symptoms. In this case, in addition to the B symptoms, the patient had splenomegaly, leukocytosis, eosinophilia, slight thrombocytopenia, anemia, elevated LDH, slightly elevated AST and ALT, and increased ALP [11]. Elevated serum ALP values are often seen in Hodgkin's lymphoma; however, elevations do not always indicate bone marrow or liver involvement. Furthermore, recent studies showed that elevated ALP is a negative prognostic indicator for overall survival in Hodgkin's lymphoma [12]. The increased ALP in Hodgkin's lymphoma is considered to be the liver isoform.

4.1.6 Patient F

A 22-year-old G1P0 female at 32 weeks' gestation presents to the emergency department in active labor. She is transferred and gives birth to a male infant weighing

2 kg. His Apgar scores are 5 at 1 min and 7 at 5 min. Owing to trouble breathing, he is intubated and taken to the neonatal intensive care unit. A chest X-ray is performed to verify placement of the endotracheal tube. The radiologist inquires about the baby's history. He says that the X-rays show little to no bone deposition in the infant. Because the infant is stable, the neonatologist goes and talks to the mother. She says that she is married to her first cousin and had little prenatal care. She says the pregnancy was uneventful. The doctor orders some laboratory tests on the mother.

The mother's laboratory tests reveal a hemoglobin of 105 g/L, hematocrit 32.1%, WBC 7.5×10^9/L, and platelets 310×10^9/L. Her comprehensive metabolic panel showed: Na 142 mmol/L, K 4.3 mmol/L, Cl 109 mmol/L, bicarbonate 27 mmol/L, AST 42 U/L, ALT 32 U/L, ALP 33 U/L, total bilirubin 24 µmol/L, creatinine 61 µmol/L, BUN 6.8 mmol/L, and glucose 6.2 mmol/L. A repeat ALP is ordered and it comes back as 29 U/L (see reference intervals in Table 4.1).

The doctor then orders laboratory test on the father. His laboratory tests show: Na 141 mmol/L, K 4.2 mmol/L, Cl 108 mmol/L, bicarbonate 27 mmol/L, AST 53 U/L, ALT 58 U/L, ALP 25 U/L, total bilirubin 20.5 µmol/L, creatinine 68.6 µmol/L, BUN 8.9 mmol/L, and glucose 5.7 mmol/L. Again the ALP is repeated and it comes back as 26 U/L.

Table 4.1: Reference range for ALP (U/L).

Age, years	Male	Female
<1	70–350	70–350
1–3	125–320	70–350
4	149–369	169–372
5	179–416	162–355
6	179–417	169–370
7	172–405	183–402
8	169–401	199–440
9	175–411	212–468
10	191–435	215–476
11	185–507	178–526
12	185–562	133–485
13	182–587	120–449
14	166–571	153–362
15	138–511	75–274
16	102–417	61–264
17	69–311	52–144
24–45	45–115	37–98
46–50	45–115	39–100
51–55	45–115	41–108
56–60	45–115	46–118
61–65	45–115	50–130
66 and older	45–115	55–142

When the baby is 3 days old, the neonatologist orders more laboratory tests including a urinary phosphoethanolamine. Again, the ALP is repeated and it comes back as 10 U/L. The urinary phosphoethanolamine is 4.42 μmol/mg of creatinine (normal is <0.33).

4.1.6.1 Discussion

This infant has congenital (infantile) hypophosphatasia, which is caused by a deficiency of tissue non-specific ALP (TNAP). If the individual is heterozygous like each of the parents, the phenotype is mild and is evident only by a decreased total ALP. However, in homozygous individuals, the phenotype can be severe and ranges from stillbirth to pathological fractures in adulthood. There are five forms of hypophosphatasia: perinatal, infantile, childhood, adult, and odontohypophosphatasia. Both the perinatal and infantile forms are autosomal recessive. The childhood form can be autosomal recessive or autosomal dominant, and the adult forms are autosomal dominant. Approximately 65 unique mutations in TNAP have been reported, most of which are missense mutations.

The infant's blood was sent for mutation analysis and came back as homozygous for R54C in exon 4, which results from a base change 211C > T [13]. The urinary phosphoethanolamine is elevated because it is an endogenous TNAP substrate, as is inorganic pyrophosphate and pyridoxal 5-phosphate [14]. In the future, this couple can benefit from genetic counseling, as prenatal diagnosis of hypophosphatasia is possible via DNA sequencing [15].

4.2 Biochemistry and physiology

4.2.1 Structure

ALP is an enzyme belonging to the hydrolase class (EC 3.1.3.1) that catalyzes removal of an orthophosphate from orthophosphoric monoesters under alkaline conditions [16]. ALP is present in almost all tissues and is in higher concentration in tissues that perform an absorptive or excretory function. Thus, the tissues with the highest ALP concentrations (greatest to least) are the placenta, small intestine, bone, kidney, and liver. There are several different forms of ALP, some of which are true isoenzymes – forms from different tissues within the same organism that catalyze the same reaction. There are four different genes that encode for the different ALP isoenzymes: one gene on chromosome 1 and three genes on chromosome 2. Chromosome 1 contains a gene that codes for TNAP, an isoenzyme found in the kidney, liver, and bones. The kidney, liver, and bone isoforms of TNAP differ in their carbohydrate content. There are four known isoforms of bone ALP: B1, B2, B/l, and B1x, which differ in activity mainly due to differences in glycosylation [17]. The genes on chromosome 2 encode for

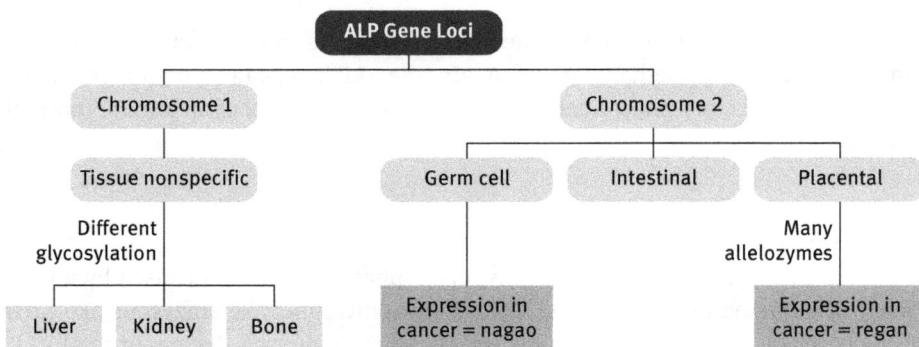

Figure 4.1: Alkaline phosphatase (ALP) gene loci. The diagram shows the four different isoforms of ALP that are encoded by genes on two chromosomes. All isoforms and isoenzymes are glycoproteins, which is a source of heterogeneity. In addition, differential cleavage or preservation of the glycosylphosphatidylinositol (GPI) anchor can lead to different isoforms. Adapted from [18].

placental ALP (PLAP), IAP, and germ cell ALP (Figure 4.1). PLAP and germ cell ALP are expressed at very low levels in the thymus and testis, and PLAP is expressed at much higher levels during pregnancy, especially during the third trimester. PLAP has many allozymes, which vary in their biochemical properties [19].

All ALP forms are glycosylphosphatidylinositol (GPI) anchored to the membrane at their carboxy terminal. Thus, the C-terminal tail is proteolyzed and the GPI anchor is added to Asp484 [20]. Phospholipases C and D cleave off the GPI anchor, thus allowing ALP to enter the blood stream [21, 22]. ALP is usually a homodimer, except under cancerous or other unusual conditions when fetal genes are derepressed and heterodimers exists [23]. Notably, placental and germ cell ALP have ~98% homology, PLAP and IAP have ~87% homology, and IAP and TNAP have only ~57% homology [18].

Several X-ray crystallographic structures of ALP are known including ALP from *Escherichia coli* [protein data bank (pdb) codes 3dyc and 2g9y] [24, 25], Antarctic bacterium TAB5 (pdb code 2iuc) [26], and human PLAP (pdb codes 1ew2, 1zeb, 1zed, 1zef, and 2glq) [27–29]. Mammalian ALPs differ from those of *E. coli* by their allosteric mechanisms and inhibition by L-amino acids such as L-Phe, L-Trp, L-homoarginine, L-Leu, and levamisole via an uncompetitive mechanism [30–32]. PLAP has lower catalytic rate (k_{cat} 460 s^{-1}) than IAP (k_{cat} 2,500 s^{-1}), or TNAP (k_{cat} 2,100 s^{-1}) [33].

PLAP comprises 484 amino acids and consists of a central β-sheet surrounded by α-helices, as shown in Figure 4.2a [27–29]. Notably, the catalytic site consisting of Ser92 requires two Zn^{2+} and one Mg^{2+} [16, 28]. Zn^{2+} is constitutively required for activity and is an integral component. However, Mg^{2+}, Mn^{2+}, Ca^{2+}, and Co^{2+} serve as activators, and cyanide, phosphate, borate, and oxalate ions serve as inhibitors. Cation binding such as Mg^{2+} or Ca^{2+} generates a conformational change that leads to enhanced enzymatic activity.

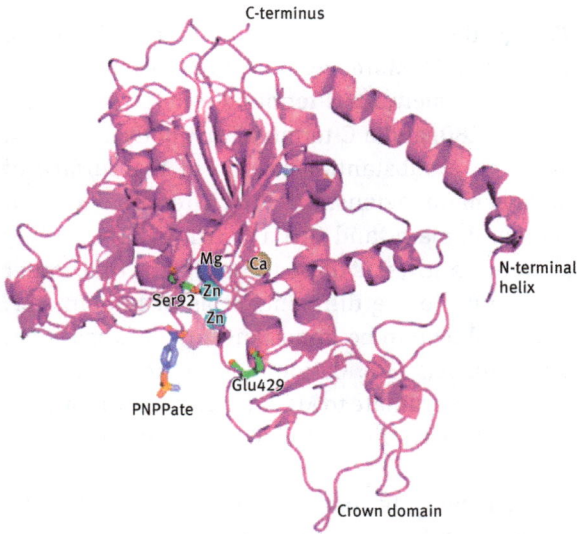

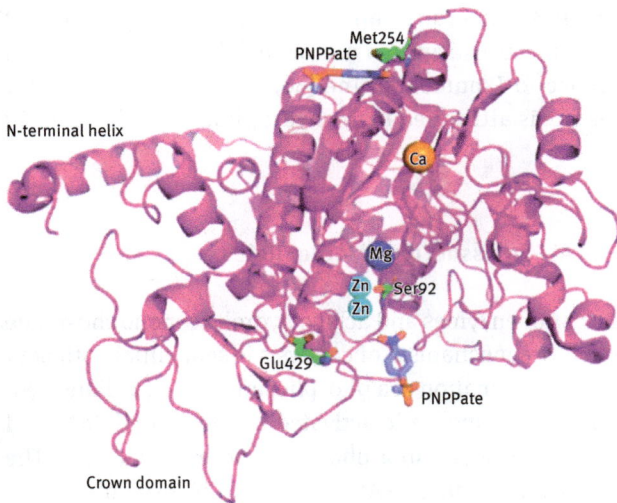

Figure 4.2: (a) Ribbon diagram of PLAP. Active site and cation binding sites in PLAP. PLAP is shown as a magenta ribbon with α-helices represented as ribbon cylinders and β-strands shown as arrows (pdb code 1zef [28]). Zn^{2+} is shown as cyan spheres, Mg^{2+} is a blue sphere, and Ca^{2+} is an orange sphere. Ser92 and Glu429 are shown as sticks and colored by atom type: carbon atoms are green, oxygen is red, and nitrogen is blue. The PNPPate inhibitor is shown in sticks and colored by atom type as above except carbon is purple. The N-terminal helix, C-terminus, and crown domain are labeled. (b) Peripheral inhibitor site of PLAP. PLAP is shown as a magenta ribbon with α-helices represented as ribbon cylinders and β-strands shown as arrows (pdb code 1zef [28]). Zn^{2+} is shown as cyan spheres, Mg^{2+} is a blue sphere, and Ca^{2+} is an orange sphere. Ser92, Met254, and Glu429 are shown as sticks and colored by atom type: carbon atoms are green, cysteines are orange, oxygen is red, and nitrogen is blue. The PNPPate inhibitor is shown in sticks and colored by atom type as above except carbon is purple. The N-terminal helix, C-terminus, and crown domain are labeled.

Compared with ALP from *E. coli*, the surface residues of human PLAP are very different with only ~8% similarity [16]. Moreover, PLAP has several additional regions not present in *E. coli* ALP, namely, an N-terminal α-helix, an α-helix and β-strand comprising residues 208–280, and a C-terminal distinctly organized β-sheet [16]. As shown in Figure 4.2a, a substantial portion of the surface of human ALP can be classified into three main regions: (i) a long N-terminal α-helix, (ii) flexible loop known as the crown domain, and (iii) a fourth cation binding site usually occupied by Ca^{2+} [16]. Importantly, it is these three regions that confer specificity for macromolecular substrate to the different forms of human ALP. The long N-terminal α-helix has been shown to be important for conformational changes that occur during enzyme catalysis [34]. Cation metal binding sites M1 and M2 coordinate to Zn^{2+}, M3 prefers to coordinate to Mg^{2+}, and M4 coordinates to Ca^{2+} [16, 27–29]. Notably, all human ALPs have five cysteines that form two disulfide bonds [35].

Germ cell ALP differs from PLAP by only one residue. Interestingly, this single amino acid change is responsible for the differences in uncompetitive inhibition of the two enzymes. PLAP is inhibited by L-Phe and germ cell ALP is inhibited by L-Leu [33]. TNAP has numerous amino acid substitutions, most of which are ionic. These hydrophilic substitutions are believed to be the cause of a switch in inhibition preference to L-homoarginine [36, 37]. The preference for hydrophobic inhibitors at this site is attributed to the hydrophobic residue Met254 (Figure 4.2b).

4.2.2 Physiological function of ALP in tissue/blood/other fluids

ALP belongs to the hydrolase class of enzymes and acts to hydrolyze monophosphate esters under alkaline conditions. The mechanism of catalysis is sequential with activation of the catalytic serine by Zn^{2+}, formation of a phosphoserine intermediate, hydrolysis of the phosphoserine by a water molecule activated by the second Zn^{2+}, and release of the phosphate product or transfer to a phosphate acceptor [38–40]. The exact function of ALP is still unclear. In bone, TNAP is produced by osteoblasts and is thought to increase bone mineralization by removal of inorganic pyrophosphates, which, if present, inhibit bone formation [41, 42]. PLAP has been found to stimulate DNA synthesis and cell proliferation in fibroblasts in the presence of Zn^{2+}, Ca^{2+}, and insulin [43].

4.2.3 Tissue sources of ALP

ALP is found in almost all tissues in the body. As seen in Figure 4.1, germ cell ALP and PLAP can both be inappropriately expressed in certain cancers leading to elevated total ALP values in various cancers [44–46]. Derepression of the *PLAP* gene in certain cancers results in an increase in PLAP; the PLAP expressed by derepression is termed the Regan isoenzyme [47]. The Regan isoenzyme is easily identified from other ALP forms due to its stability at 65°C [47]. As shown in Figure 4.1, inappropriate expression of modified germ cell ALP by tumors is termed the Nagao isoenzyme [48]. Inappropriate expression of PLAP in a cell that also makes IAP where heterodimerization can occur is termed the Kasahara isoenzyme [23, 49]. The Regan, Nagao, and Kasahara isoenzymes are all believed to result from tumor derepression of the fetal *PLAP* or germ cell *ALP* genes.

ALP is usually measured in serum; however, a role exists for measuring PLAP in cerebral spinal fluid (CSF). In serum and CSF, PLAP (Regan isoenzyme) can be used as a tumor marker for germ cell tumors [44].

4.2.4 Clearance/metabolism of enzyme

The half-life of the ALP isoforms varies. PLAP has a half-life of 7 days [50], the liver form has a half-life of 3 days, the bone form has a half-life of 1 to 2 days [51], and IAP has a half-life of minutes [52]. The day-to-day variability of total ALP in blood ranges from 5% to 10% for total ALP, with bone ALP having a variability of ~20% [53]. Owing to the faster clearance of desialyated ALP forms, it is hypothesized that ALP may be desialyated prior to elimination [54].

4.2.5 Reference ranges

Reference intervals for ALP vary based on the methodology used for measuring the enzyme. The reference intervals for total serum ALP also vary based on age and gender (Tables 4.1 and 4.2). Because bone ALP is made by osteoblasts, which are active during periods of bone growth, children and adolescents have higher ALP ranges. Moreover, because girls experience their growth spurt in puberty earlier than boys, their ALP

Table 4.2: Reference ranges for ALP isoenzymes (U/L) using gel analyses.

Age, years	Liver 1	Liver 2	Bone	Intestine
0–6	7.0–112.7	3.0–41.5	43.5–208.1	0.0–37.7
7–9	7.4–109.1	4.0–35.6	41.0–258.3	0.0–45.6
10–13	7.8–87.6	3.3–37.8	39.4–346.1	0.0–40.0
14–15	10.3–75.6	2.2–32.1	36.4–320.5	0.0–26.4
16–18	13.7–78.5	1.4–19.7	32.7–214.6	0.0–12.7
19 and older	16.2–70.2	0.0–5.8	12.1–42.7	0.0–11.0

Table 4.3: Reference ranges for ALP during pregnancy[1].

Gestation stage	Total serum ALP, U/L[1]
7–17 weeks	35–80
17–24 weeks	39–105
24–28 weeks	46–115
28–31 weeks	53–119
31–34 weeks	66–177
34–38 weeks	87–228
Predelivery	98–296
Postpartum	52–164

Note: 1. Adapted from [6].

normal range increases at an earlier age (Table 4.1). Total ALP values among males and females stabilize after 20 years of age and remain similar until women enter menopause and begin losing bone due to the decrease in estrogen. Thus, the ALP reference range is higher for women over 50 years old. Furthermore, total ALP ranges are 10–15% higher for African-American men and women [55]. The usual distribution of ALP in adult serum is ~40% liver and ~60% bone.

During pregnancy, PLAP increases, particularly in the third trimester, and results in a two- to three-fold increase in total ALP [56]. Thus, the reference ranges for total serum ALP are different during pregnancy as shown in Table 4.3. Furthermore, an increased BMI has been shown to increase ALP by ~10% [57] and smoking has also been shown to increase ALP by ~10% due to production of placental-like ALP by the lung [58]. Oral contraceptives that contain estrogen are associated with decreased ALP [59], whereas antiepileptic drugs such as phenobarbital or phenytoin cause an increase in ALP usually via an increase in liver ALP [60]. Furthermore, individuals on cardiopulmonary bypass, receiving transfusions, or undergoing pheresis, where citrate is used to keep the blood anticoagulated, will have lower ALP activity in their blood. The low ALP value is attributable to chelation of cations needed by ALP for activity by citrate. Accordingly, anticoagulant tubes containing citrate, oxalate, or potassium EDTA are inappropriate for measuring ALP in blood.

4.3 Chemical pathology

4.3.1 Liver disease

ALP increases in liver disease or injury such as hepatitis [61–64]. The most substantial increases occur with biliary tree obstruction or hepatic cancer/hepatic metastasis [61, 65]. Liver cirrhosis and other forms of hepatic disease or injury can result in increased ALP due to damage and impairment of the hepatic asialoglycoprotein receptors, which are responsible for clearing IAP [66]. ALP is increased more in extrahepatic obstruction compared with intrahepatic obstruction. Notably, liver injury causes an increase in the release of liver ALP from the cell membrane; however, obstructive hepatobiliary disease results in an increase in bile salts which dissolve the cell membrane and releases ALP bound to lipoproteins from the cell membrane. This results in two sizes of liver ALP that is released, termed liver 1 and liver 2 [67].

4.3.2 Bone disease

ALP values are increased in numerous bone diseases where osteoblast activity is increased. Osteoblasts build bone and also produce bone ALP. Bone ALP concentrations in blood are increased 10–25 times the upper limit in Paget's disease [5, 68]. Bone ALP values are also substantially increased in osteogenic bone cancer [69]. Moreover, bone ALP values are increased two to four times the upper limit in osteomalacia and rickets, which are caused by vitamin D deficiency and have an elevated rate of bone breakdown and remodeling by osteoclasts and osteoblasts. Importantly, children who are actively growing have elevated levels of bone ALP and therefore elevated total ALP; hence, the normal reference ranges of total serum ALP are adjusted for age and sex as girls have their growth spurt earlier compared with boys (Tables 4.1 and 4.2).

4.3.3 Metastatic cancer

Total ALP increases in cancers that metastasize to the liver or to the bone. The increase in ALP reflects the site(s) of metastases with either bone or liver ALP (or both) being increased. Approximately 70% of patients with advanced prostate or breast cancer develop bone metastases and in approximately 40% of patients with solid cancers develop bone metastases [70]. The median bone ALP in prostate cancer patients with bone metastases was found to be 38.5 µg/L (normal men = 3.7–20.9 µg/L) [71]. Moreover, the median bone ALP in breast cancer patients with bone metastases was found to be 45.7 µg/L (normal for premenopausal women = 2.9–14.5 µg/L and postmenopausal women = 3.8–22.6 µg/L) [71]. In patients with hepatomas, the tumor itself releases ALP into the blood causing an increase in ALP [72].

4.3.4 Miscellaneous pancreatic disorders

Pancreatic ductal adenocarcinoma or other pancreatic cancers that cause biliary obstruction can result in very large increases in total ALP [73]. Other pancreatic cancers that do not cause blockage of the biliary tract will cause anywhere from a moderate increase to no increase in ALP values. Pancreatic insufficiency can cause an increase in ALP and is usually seen in individuals with chronic pancreatitis. The fat soluble vitamins A, D, E, and K are dependent on the exocrine function of the pancreas for absorption. Hence, with pancreatic insufficiency, vitamins A, D, E, and K are poorly absorbed resulting in low levels of vitamin D3 [74]. This leads to a decreased 1,25-dihydroxyvitamin D level and decreased bone density. The decrease in bone density will activate osteoblasts and increase bone ALP causing an increase in total ALP.

4.3.5 Cystic fibrosis

In patients with cystic fibrosis, increased ALP values can be associated with hepatobiliary disease and complications [48]. Adult cystic fibrosis patients have increased bone turnover and can have elevated bone ALP. This is in contrast to the elevated ALP seen in hepatobiliary complications, which is largely attributable to liver ALP forms.

4.3.6 Chronic renal failure

Chronic renal failure (CRF) and end-stage renal disease (ESRD) are associated with abnormal bone turnover and abnormal mineralization. When the glomerular filtration rate falls below 60 mL/min, the kidney starts to lose its ability to excrete phosphate. Thus, the plasma phosphate level increases, which in turn causes an increase in parathyroid hormone (PTH) and a concomitant decrease in 1,25-dihydroxyvitamin D levels [75, 76]. Ultimately, CRF and ESRD can lead to high-turnover bone disease, mixed osteodystrophy, and/or low-turnover bone disease, all of which are associated with increased ALP [77]. The main increase in ALP in CRF/ESRD is attributable to an increase in the bone forms of ALP; however, the other forms of ALP can also be increased.

Following renal transplant, bone mass is lost rapidly particularly within the first 6 months [78]. This loss in bone density is largely attributable to the use of high doses of steroids employed to prevent rejection. The steroids cause decreased osteoblastogenesis, increased osteoblast and osteocyte apoptosis, and suppressed osteoblast bone formation [78, 79]. Thus, following renal transplant ALP will decrease due to the drop in the bone-specific forms.

4.3.7 Drug therapy

Increases in liver ALP can be seen as a reaction to various drugs such as cortisone, chlorpromazine, allopurinol, methyldopa, propranolol, tricyclic antidepressants, barbiturates, phenytoin, cimetidine, furosemide, halothane, papaverine, birth control pills containing estrogen, statins, and alcohol. Drugs that decrease ALP include clofibrate and sulfonamides, [53, 59].

4.3.8 Miscellaneous

Individuals with both primary and secondary hyperparathyroidism have elevated total ALP due to the increase in bone turnover and bone ALP production by osteoblasts [80]. In primary hyperparathyroidism, the parathyroid gland makes excess PTH, increasing bone resorption and raising the serum calcium. In secondary hyperparathyroidism, PTH is released from the parathyroid in response to low calcium (as occurs in CRF) and bone resorption increases along with ALP.

During the third trimester of pregnancy, the total serum ALP value can double due to the contribution of PLAP [56]. Also, individuals older than 50 years of age, particularly postmenopausal women, may have a total ALP level 1.5 times the upper limit of normal [81, 82]. Vitamin D deficiency and resultant osteopenia/osteoporosis is associated with elevated bone turnover and increased ALP [83]. Nutritional deficiencies such as protein deficiency, magnesium deficiency, poor nutrition, and vitamin C deficiency, as well as vitamin D excess can cause the total ALP value to be decreased or low.

ALP is typically higher in individuals with blood types O and B [83]. IAP is released from the cell membrane into duodenal fluid following ingestion of food. The IAP then enters lymphatic fluid where it is picked up by the red blood cells. In blood type O and B individuals serum ALP has been shown to increase by up to 30 U/L following a meal [84].

ALP also increases in various diseases such as Crohn's disease, ulcerative colitis, peptic and/or duodenal ulcer, splenic infarct, non-alcoholic steatohepatitis, and small bowel infarction [61–64].

4.4 Analysis

4.4.1 Technical problems

4.4.1.1 Specimen

The type of vacutainer collection tube used to collect a sample for ALP analysis can affect the results. Tubes containing citrate or EDTA, which bind cations decrease ALP activity because ALP requires Zn^{2+} for its activity. In addition, specimens collected in sodium fluoride and potassium oxalate are also unacceptable. Serum or heparinized plasma samples are preferred for ALP measurement.

4.4.1.2 Analyte stability

ALP in serum has been shown to be stable for 56 h at room temperature [85], 1 week refrigerated, and 1 year frozen [59]. With refrigeration, ALP activity will increase 2% per day [59].

In chromogenic measurements of total serum ALP, the activity of ALP is enhanced if a phosphate-accepting buffer is used. Examples of buffers that facilitate ALP activity are: AMP, diethanolamine, tris(hydroxymethyl)aminomethane, ethylaminoethanol, and N-methyl-D-glucosamine. Various methods have been used to isolate or analyze the various forms of ALP. Levimasole inhibits bone and liver forms of ALP, and phenylalanine inhibits placental and intestinal forms of ALP. Notably, assays with phenylalanine and levamisole are poorly reproducible. Heat inactivation has been used because the placental and germ cell forms are very heat stable and bone ALP is very heat labile, with the liver form having intermediate heat stability.

4.4.1.3 Test interferences

Chelators of Zn^{2+} and Mg^{2+} such as EDTA, citrate, and oxalate interfere with the measurement of ALP by reducing the activity of ALP. With the addition of exogenous Zn^{2+} and Mg^{2+}, the chromogenic assay of ALP is very robust with minimal interferences. Notably, samples that are extremely hemolyzed, icteric, or lipemic can interfere with ALP measurement via their effect on absorbance measurements taken at 405 nm.

4.4.2 Reference method

The reference method for ALP measurement was described by Tietz et al. in 1983. In the reference method, ALP converts p-nitrophenyl phosphate to p-nitrophenoxide and phosphate, with the buffer 2A2M1P acting as a phosphate acceptor. The yellow-colored p-nitrophenoxide is measured at 405 nm [86].

4.4.3 Isoenzyme analysis

A variety of methods have been used to measure the activity of the different ALP isoenzymes.

4.4.3.1 Heat inactivation

As shown in Table 4.4, by incubating serum samples at 65°C for 30 min, the bone ALP forms are readily inactivated ("bone burns") and the liver forms lose some (~40%) but not all activity. PLAP and germ cell ALP retain activity following heating [87, 88].

Table 4.4: Chemical properties of ALP isoenzymes[1].

Properties	Liver 1	Liver 2	Bone	Placenta	Intestinal	Kasahara	Nagao	Regan
Heat stability	Unstable	Unstable	Very unstable	Stable	Unstable	Stable	Very stable	Very stable
L-Phe inhibition	Weak	Weak	Weak	Strong	Strong	Moderate	Strong	Strong
L-Homoarginine inhibition	Moderate	Strong	Strong	Weak	Weak	Weak	Weak	Weak
L-Leu inhibition	Weak	Moderate	Moderate	Weak	Weak	Moderate	Moderate	Weak
Urea inactivation	Strong	Strong	Very strong	Moderate	Moderate			Strong
Electrophoretic mobility	α1	α2	α2	α2/β	β	α1	α2/β	α2/β

Note: 1. Adapted from [89].

4.4.3.2 Gel electrophoresis analysis

ALP isoenzymes in serum can also be assessed by gel electrophoresis at pH 9.1 [90]. Bone and liver isoenzymes have identical amino acid sequence and electrophoresis requires some modifications to allow their separation. Advantage is taken by the fact that liver and bone isoenzymes differ in sialation. Samples are analyzed twice, untreated and then treated by reacting the sample with wheat germ lectin that reacts with the bone isoenzyme, which has a much greater number of sialic acids, and precipitates near the point of application.

Another modification that can be used to separate liver and bone isoenzymes is neuraminidase treatment, which removes terminal sialic acid moieties. Serum is incubated at 37°C for 15 min prior to application on the gel. This allows for better separation because the bone ALP sialic acid moieties are more susceptible to neuraminidase, resulting in decreased electrophoretic mobility. In addition, IAP can be readily identified by treating the serum sample with neuraminidase overnight and then subjecting the sample to electrophoresis. Neuraminidase removes terminal sialic acid moieties from all forms of ALP except IAP because IAP does not have any terminal sialic acid moieties. Following electrophoresis, the isoenzymes can be visualized using 5-bromo-4-chloro-3-indolyl phosphate/nitro blue tetrazolium in aminomethyl propanol buffer, pH 10.0. As shown in Figure 4.3, liver ALP migrates the most quickly towards the anode followed by bone ALP, then IAP, and lastly PLAP. Notably, the high-molecular weight form of liver ALP (liver 1) that is released when bile salts accumulate that was described previously migrates anodally to normal sized liver ALP (liver 2) on cellulose acetate, which is depicted in Figure 4.3. However, the liver 1 form migrates very slowly in starch and polyacrylamide gels and is thus not seen migrating anodal to the liver 2 form.

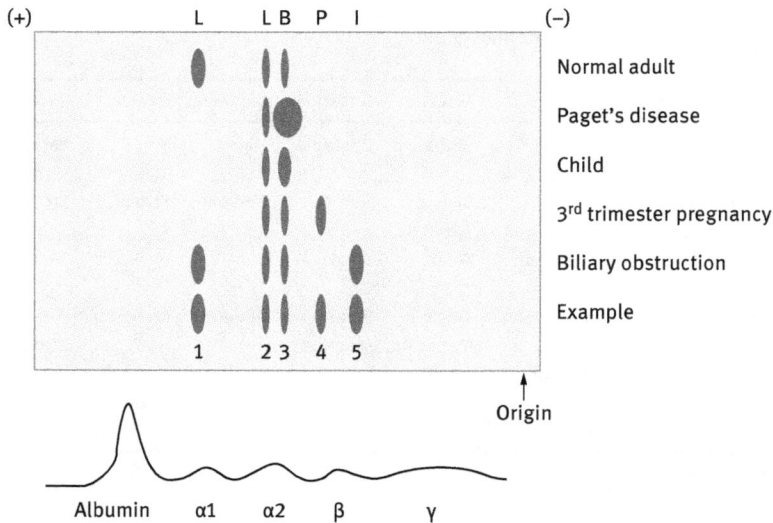

Figure 4.3: Electrophoretic pattern of ALP isoenzymes and isoforms on cellulose acetate. The pattern seen in normal adults is shown on top with bands 1, 2, and 3. Paget's disease shows the elevation of bone-specific ALP (band 3). Children have an increased bone-specific ALP due to bone growth as shown in roll 3 with an increase in band 3. The third trimester of pregnancy is marked by an increase in PLAP (band 4), and biliary obstruction shows all forms except placental. Lastly, the example, band 1 is liver 1, band 2 is liver 2, band 3 is bone, band 4 is placental, and band 5 is intestinal. A characteristic densitometer diagram for serum protein electrophoresis is shown below the gel diagram.

4.4.3.3 Chemical inactivation

Various compounds act to inhibit certain isoenzymes of ALP. PLAP and IAP are inhibited by L-Phe and germ cell ALP is inhibited by L-Leu [33]. Homoarginine acts to inhibit liver and bone ALP [91]. Levamisole reversibly and noncompetitively inhibits liver and bone ALP but not IAP and PLAP [31]. Theophylline noncompetitively inhibits liver ALP and IAP but not PLAP [92].

Forsman and O'Brien [93] described a method using urea and L-Phe to measure ALP isoenzymes. In this method, three separate ALP analyses are performed: total ALP, L-Phe inhibition, and urea inactivation. The PLAP and IAP forms are inhibited by L-Phe, and urea causes inactivation of the bone and liver forms. The urea-treated ALP activity values (two-component decay) are resolved via a linear equation obtained from known quantities of bone and liver ALP that were treated with urea.

4.4.3.4 Immunoassay

Bone ALP can be measured using immunoassay. Patient serum is mixed with a mouse monoclonal antibody specific for bone ALP. Liver ALP has some cross-reactivity (6–20%) in this assay and can cause substantial interference when liver ALP is increased [94, 95].

4.5 Questions and answers

1. Which of the following ALP isoenzymes retains activity following heating for
    10 min as 65°C?
    (a) Regan
    (b) Liver
    (c) Bone
    (d) Intestinal
    (e) Prostate
2. Which of the following conditions results in the greatest increase in serum ALP
    activity?
    (a) Pregnancy
    (b) Rickets
    (c) Liver cirrhosis
    (d) Paget's disease
    (e) Osteomalacia
3. Which ALP isoenzyme exhibits the shortest half-life?
    (a) Placental ALP
    (b) Bone ALP, with a half-life of 12 h
    (c) Regan isoenzyme
    (d) Intestinal ALP with a half-life of 8 h
    (e) Intestinal ALP with a half-life of less than 30 min
4. Which of the following is required for ALP activity?
    (a) Ca^{2+}
    (b) Zn^{2+}
    (c) NADH
    (d) NAD^+
    (e) PO_4^{2+}
5. Which of the following ALP isoenzymes is most readily deactivated by heating at
    65°C?
    (a) Intestinal
    (b) Liver
    (c) Bone
    (d) Regan
    (e) Placental

Acknowledgements

I would like to thank Dr Mitch Scott for his help in editing this chapter.

References

[1] Arulventhan R, Larcos G, Gruenewald SM, Farrell GC. Primary sclerosing cholangitis: correlation of hepatobiliary scintigraphy with clinical and laboratory status. Australas Radio 2006,50,330–4.

[2] Broomé U, Bergquist A. Primary sclerosing cholangitis, inflammatory bowel disease, and colon cancer. Semin Liver Dis 2006,26,31–41.

[3] Silveira MG, Lindor KD. Primary sclerosing cholangitis. Can J Gastroenterol 2008,22,689–98.

[4] Magnusson P, Davie MW, Sharp CA. Circulating and tissue-derived isoforms of bone alkaline phosphatase in Paget's disease of bone. Scand J Clin Lab Invest 2010,70,128–35.

[5] Varenna M, Zucchi F, Galli L, Manara M, De Marco G, Sinigaglia L. Demographic and clinical features related to a symptomatic onset of Paget's disease of bone. J Rheumatol 2010,37,155–60.

[6] Larsson A, Palm M, Hansson LO, Axelsson O. Reference values for clinical chemistry tests during normal pregnancy. BJOG 2008,115,874–81.

[7] Jamjute P, Ahmad A, Ghosh T, Banfield P. Liver function test and pregnancy. J Matern Fetal Neonatal Med 2009,22,274–83.

[8] Golik A, Rubio A, Weintraub M, Byrne L. Elevated serum liver enzymes in obesity: a dilemma during clinical trials. Int J Obes 1991,15,797–801.

[9] Ognibene A, Pala L, Messeri G, Rotella CM, Berti P. Relations between intestinal alkaline phosphatase activity and insulin secretion in obese patients. Clin Chem 1997,43,1672–3.

[10] Hansen GH, Niels-Christiansen LL, Immerdal L, Nystrøm BT, Danielsen EM. Intestinal alkaline phosphatase: selective endocytosis from the enterocyte brush border during fat absorption. Am J Physiol Gastrointest Liver Physiol 2007,293,G1325–32.

[11] Aisenberg AC, Kaplan MM, Rieder SV, Goldman JM. Serum alkaline phosphatase at the onset of Hodgkin's disease. Cancer 1970,26,318–26.

[12] Itoh K, Kinoshita T, Watanabe T, Yoshimura K, Okamoto R, Chou T, et al. Prognostic analysis and a new risk model for Hodgkin lymphoma in Japan. Int J Hematol 2010,91,446–55.

[13] Henthorn PS, Raducha M, Fedde KN, Lafferty MA, Whyte MP. Different missense mutations at the tissue-nonspecific alkaline phosphatase gene locus in autosomal recessively inherited forms of mild and severe hypophosphatasia. Proc Natl Acad Sci USA 1992,89,9924–8.

[14] Mornet E. Hypophosphatasia: the mutations in the tissue-nonspecific alkaline phosphatase gene. Hum Mutat 2000,15,309–15.

[15] Simon-Bouy B, Taillandier A, Fauvert D, Brun-Heath I, Serre JL, Armengod CG, et al. Hypophosphatasia: molecular testing of 19 prenatal cases and discussion about genetic counseling. Prenat Diagn 2008,28,993–8.

[16] Millán JL. Alkaline phosphatases: structure, substrate specificity and functional relatedness to other members of a large superfamily of enzymes. Purinergic Signal 2006,2,335–41.

[17] Halling Linder C, Narisawa S, Millán JL, Magnusson P. Glycosylation differences contribute to distinct catalytic properties among bone alkaline phosphatase isoforms. Bone 2009,45, 987–93.

[18] Moss DW. Perspectives in alkaline phosphatase research. Clin Chem 1992,38,2486–92.

[19] Wennberg C, Kozlenkov A, Di Mauro S, Fröhlander N, Beckman L, Hoylaerts MF, et al. Structure, genomic DNA typing, and kinetic characterization of the D allozyme of placental alkaline phosphatase (PLAP/ALPP). Hum Mutat 2002,19,258–67.

[20] Micanovic R, Gerber LD, Berger J, Kodukula K, Udenfriend S. Selectivity of the cleavage/ attachment site of phosphatidylinositol-glycan-anchored membrane proteins determined by site-specific mutagenesis at Asp-484 of placental alkaline phosphatase. Proc Natl Acad Sci USA 1990,87,157–61.

[21] Martini CN, Vaena de Avalos SG, del Carmen Vila M. ACTH stimulates the release of alkaline phosphatase through Gi-mediated activation of a phospholipase C and the release of inositol-phosphoglycan. Mol Cell Biochem 2004,258,191–9.

[22] Moon YG, Lee HJ, Kim MR, Myung PK, Park SY, Sok DE. Conversion of glycosylphosphatidylinositol (GPI)-anchored alkaline phosphatase by GPI-PLD. Arch Pharm Res 1999,22,249–54.

[23] Endo T, Higashino K, Hada T, Imanishi H, Muratani K, Kochibe N, et al. Structures of the asparagine-linked oligosaccharides of an alkaline phosphatase, Kasahara isozyme, purified from FL amnion cells. Cancer Res 1990,50,1079–84.

[24] Zalatan JG, Fenn TD, Herschlag D. Comparative enzymology in the alkaline phosphatase superfamily to determine the catalytic role of an active-site metal ion. J Mol Biol 2008,384,1174–89.

[25] Wang J, Kantrowitz ER. Trapping the tetrahedral intermediate in the alkaline phosphatase reaction by substitution of the active site serine with threonine. Protein Sci 2006,15, 2395–401.

[26] Wang E, Koutsioulis D, Leiros HK, Andersen OA, Bouriotis V, Hough E, et al. Crystal structure of alkaline phosphatase from the Antarctic bacterium TAB5. J Mol Biol 2006,366,1318–31.

[27] Le Du MH, Stigbrand T, Taussig MJ, Menez A, Stura EA. Crystal structure of alkaline phosphatase from human placenta at 1.8 A resolution: implication for a substrate specificity. J Biol Chem 2000,276,9158–65.

[28] Llinas P, Stura EA, Menez A, Kiss Z, Stigbrand T, Millan JL, et al. Structural studies of human placental alkaline phosphatase in complex with functional ligands. J Mol Biol 2005,350, 441–51.

[29] Llinas P, Masella M, Stigbrand T, Menez A, Stura EA, Le Du MH. Structural studies of human alkaline phosphatase in complex with strontium: implication for its secondary effect in bones. Protein Sci 2006,15,1691–700.

[30] Fishman WH, Sie HG. Organ-specific inhibition of human alkaline phosphatase isoenzymes of liver, bone, intestine and placenta; L-phenylalanine, L-tryptophan and L homoarginine. Enzymologia 1971,41,141–67.

[31] Van Belle H. Alkaline phosphatase. I. Kinetics and inhibition by levamisole of purified isoenzymes from humans. Clin Chem 1976,22,972–6.

[32] Doellgast GJ, Fishman WH. Inhibition of human placental-type alkaline phosphatase variants by peptides containing L-leucine. Clin Chim Acta 1977,75,449–54.

[33] Hummer C, Millán JL. Gly429 is the major determinant of uncompetitive inhibition of human germ cell alkaline phosphatase by L-leucine. Biochem J 1991,274,91–5.

[34] Hoylaerts MF, Ding L, Narisawa S, Van Kerckhoven S, Millan JL. Mammalian alkaline phosphatase catalysis requires active site structure stabilization via the N-terminal amino acid microenvironment. Biochemistry 2006,45,9756–66.

[35] Kozlenkov A, Manes T, Hoylaerts MF, Millán JL. Function assignment to conserved residues in mammalian alkaline phosphatases. J Biol Chem 2002,277,22992–9.

[36] Le Du MH, Millan JL. Structural evidence of functional divergence in human alkaline phosphatases. J Biol Chem 2002,277,49808–14.

[37] Kozlenkov A, Le Du MH, Cuniasse P, Ny T, Hoylaerts MF, Millán JL. Residues determining the binding specificity of uncompetitive inhibitors to tissue-nonspecific alkaline phosphatase. J Bone Miner Res 2004,19,1862–72.

[38] Xu X, Kantrowitz ER. A water-mediated salt link in the catalytic site of *Escherichia coli* alkaline phosphatase may influence activity. Biochemistry 1991,30,7789–96.

[39] Janeway CM, Xu X, Murphy JE, Chaidaroglou A, Kantrowitz ER. Magnesium in the active site of *Escherichia coli* alkaline phosphatase is important for both structural stabilization and catalysis. Biochemistry 1993,32,1601–9.

[40] Murphy JE, Kantrowitz ER. Why are mammalian alkaline phosphatases much more active than bacterial alkaline phosphatases? Mol Microbiol 1994,12,351–7.

[41] Orimo H. The mechanism of mineralization and the role of alkaline phosphatase in health and disease. J Nippon Med Sch 2010,77,4–12.

[42] Polewski MD, Johnson KA, Foster M, Millán JL, Terkeltaub R. Inorganic pyrophosphatase induces type I collagen in osteoblasts. Bone 2010,46,81–90.
[43] She QB, Mukherjee JJ, Huang JS, Crilly KS, Kiss Z. Growth factor-like effects of placental alkaline phosphatase in human fetus and mouse embryo fibroblasts. FEBS Lett 2000,469,163–7.
[44] Millán JL, Fishman WH. Biology of human alkaline phosphatases with special reference to cancer. Crit Rev Clin Lab Sci 1995,32,1–39.
[45] Iczkowski KA, Butler SL. New immunohistochemical markers in testicular tumors. Anal Quant Cytol Histol 2006,28,181–7.
[46] Sato K, Takeuchi H, Kubota T. Pathology of intracranial germ cell tumors. Prog Neurol Surg 2009,23,59–75.
[47] Nathanson L, Fishman WH. New observations on the Regan isoenzyme of alkaline phosphatase in cancer patients. Cancer 1971,27,1388–97.
[48] Crofton PM. Biochemistry of alkaline phosphatase isoenzymes. Crit Rev Clin Lab Sci 1982,16,161–94.
[49] Higashino K, Muratani K, Hada T, Imanishi H, Amuro Y, Yamamoto Y, et al. Purification and some properties of the fast migrating alkaline phosphatase in FL-amnion cells (the Kasahara isoenzyme) and its cDNA cloning. Clin Chim Acta 1990,86,151–64.
[50] Clubb JS, Neale FC, Posen S. The behavior of infused human placental alkaline phosphatase in human subjects. J Lab Clin Med 1965,66,493–507.
[51] Posen S, Grunstein HS. Turnover rate of skeletal alkaline phosphatase in humans. Clin Chem 1982,28,153–4.
[52] Kleerekoper M, Horne M, Cornish CJ, Posen S. Serum alkaline phosphatase after fat ingestion, an immunological study. Clin Sci 1970,38,339–45.
[53] McPherson RA, Pincus MR, eds. Henry's Clinical Diagnosis and Management by Laboratory methods. 21st ed. St Louis, MO, USA, Saunders Elsevier, 2007,255–6.
[54] Blom E, Ali MM, Mortensen B, Huseby NE. Elimination of alkaline phosphatases from circulation by the galactose receptor. Different isoforms are cleared at various rates. Clin Chim Acta 1998,270,125–37.
[55] Manolio TA, Burke GL, Savage PJ, Jacobs DR Jr, Sidney S, Wagenknecht LE, et al. Sex- and race-related differences in liver-associated serum chemistry tests in young adults in the CARDIA study. Clin Chem 1992,38,1853–9.
[56] Valenzuela GJ, Munson LA, Tarbaux NM, Farley JR. Time-dependent changes in bone, placental, intestinal, and hepatic alkaline phosphatase activities in serum during human pregnancy. Clin Chem 1987,33,1801–6.
[57] Salvaggio A, Periti M, Miano L, Tavanelli M, Marzorati D. Body mass index and liver enzyme activity in serum. Clin Chem 1991,37,720–3.
[58] Kallioniemi OP, Nieminen MM, Lehtinen J, Veneskoski T, Koivula T. Increased serum placental-like alkaline phosphatase activity in smokers originates from the lungs. Eur J Respir Dis 1987,71,170–6.
[59] Burtis CA, Ashwood ER, Bruns DE, eds. Tietz Fundamentals of Clinical Chemistry. St Louis, MO, USA, Sanders Elsevier, 2008,325–7.
[60] Nijhawan R, Wierzbicki AS, Tozer R, Lascelles PT, Patsalos PN. Antiepileptic drugs, hepatic enzyme induction and raised serum alkaline phosphatase isoenzymes. Int J Clin Pharmacol Res 1990,10,319–23.
[61] Wolf PL. Clinical significance of an increased or decreased serum alkaline phosphatase level. Arch Pathol Lab Med 1978,102,497–501.
[62] Qirbi AA, Moss DW. Incidence and some properties of an electrophoretically slow form of alkaline phosphatase in sera of patients with diseases of the intestine. Clin Chim Acta 1975,60,1–6.
[63] Chopra S, Griffin PH. Laboratory tests and diagnostic procedures in evaluation of liver disease. Am J Med 1985,79,221–30.

[64] Neuschwander-Tetri BA, Clark JM, Bass NM, Van Natta ML, Unalp-Arida A, Tonascia J, et al. NASH Clinical Research Network. Clinical, laboratory and histological associations in adults with nonalcoholic fatty liver disease. Hepatology 2010,52,913–24.

[65] Winkelman J, Nadler S, Demetriou J, Pileggi VJ. The clinical usefulness of alkaline phosphatase isoenzyme determinations. Am J Clin Pathol 1972,57,625–34.

[66] Sugahara K, Togashi H, Takahashi K, Onodera Y, Sanjo M, Misawa K, et al. Separate analysis of asialoglycoprotein receptors in the right and left hepatic lobes using Tc-GSA SPECT. Hepatology 2003,38,1401–9.

[67] Wolf PL. Clinical significance of serum high-molecular-mass alkaline phosphatase, alkaline phosphatase-lipoprotein-X complex, and intestinal variant alkaline phosphatase. J Clin Lab Anal 1994,8,172–6.

[68] Colina M, La Corte R, De Leonardis F, Trotta F. Paget's disease of bone: a review. Rheumatol Int 2008,28,1069–75.

[69] Demers LM, Costa L, Chinchilli VM, Gaydos L, Curley E, Lipton A. Biochemical markers of bone turnover in patients with metastatic bone disease. Clin Chem 1995,41,1489–94.

[70] Coleman RE. Skeletal complications of malignancy. Cancer 1997,80,1588–94.

[71] Garnero P, Buchs N, Zekri J, Rizzoli R, Coleman RE, Delmas PD. Markers of bone turnover for the management of patients with bone metastases from prostate cancer. Br J Cancer 2000,82,858–64.

[72] McComb RB, Bowers GN, Posen S. Alkaline Phosphatases. New York, NY, USA, Plenum Publishing Corporation, 1979.

[73] Stocken DD, Hassan AB, Altman DG, Billingham LJ, Bramhall SR, Johnson PJ, et al. Modelling prognostic factors in advanced pancreatic cancer. Br J Cancer 2008,99,883–93.

[74] Mann ST, Stracke H, Lange U, Klör HU, Teichmann J. Alterations of bone mineral density and bone metabolism in patients with various grades of chronic pancreatitis. Metabolism 2003,52,579–85.

[75] Fajtova VT, Sayegh MH, Hickey N, Aliabadi P, Lazarus JM, LeBoff MS. Intact parathyroid hormone levels in renal insufficiency. Calcif Tissue Int 1995,57,329–35.

[76] Martinez I, Saracho R, Montenegro J, Llach F. The importance of dietary calcium and phosphorous in the secondary hyperparathyroidism of patients with early renal failure. Am J Kidney Dis 1997,29,496–502.

[77] Elder G. Pathophysiology and recent advances in the management of renal osteodystrophy. J Bone Miner Res 2002,17,2094–105.

[78] Julian BA, Laskow DA, Dubovsky J, Dubovsky EV, Curtis JJ, Quarles LD. Rapid loss of vertebral mineral density after renal transplantation. N Engl J Med 1991,25,544–50.

[79] Weinstein RS, Jilka RL, Parfitt AM, Manolagas SC. Inhibition of osteoblastogenesis and promotion of apoptosis of osteoblasts and osteocytes by glucocorticoids. Potential mechanisms of their deleterious effects on bone. J Clin Invest 1998,102,274–82.

[80] Lind L, Wengle B, Lithell H, Ljunghall S. Reduction in serum alkaline phosphatase levels by treatment with active vitamin D (alphacalcidol) in primary and secondary hyperparathyroidism and in euparathyroid individuals. Scand J Urol Nephrol 1991,25,233–6.

[81] Schiele F, Henny J, Hitz J, Petitclerc C, Gueguen R, Siest G. Total bone and liver alkaline phosphatases in plasma: biological variations and reference limits. Clin Chem 1983, 29,634–41.

[82] Majkić-Singh N, Ilić M, Ignjatović S, Aleksandra-Postić-Grujin. Assessment of four biochemical markers of bone metabolism in postmenopausal osteoporosis. Clin Lab 2002,48,407–13.

[83] Agbedana EO, Yeldu MH. Serum total, heat and urea stable alkaline phosphatase activities in relation to ABO blood groups and secretor phenotypes. Afr J Med Med Sci 1996,25,327–9.

[84] Domar U, Karpe F, Hamsten A, Stigbrand T, Olivecrona T. Human intestinal alkaline phosphatase – release to the blood is linked to lipid absorption, but removal from the blood is not linked to lipoprotein clearance. Eur J Clin Invest 1993,23,753–60.

[85] Boyanton BL Jr, Blick KE. Stability studies of 24 analytes in human plasma and serum. Clin Chem 2002,48,2242–7.
[86] Tietz NW, Burtis CA, Duncan P, Ervin K, Petitclerc CJ, Riker AD, et al. A reference method for measurement of alkaline phosphatase activity in human serum. Clin Chem 1983,29,751–61.
[87] Posen S, Neale FC, Clubb JS. Heat inactivation in the study of human alkaline phosphatases. Ann Intern Med 1965,62,1234–43.
[88] Neale FC, Clubb JS, Hotchkis D, Posen S. Heat stability of human placental alkaline phosphatase. J Clin Pathol 1965,18,359–63.
[89] Nemesanszky E. Alkaline phosphatase. In: Lott JA, Wolf PL, eds. Clinical Enzymology. A Case Oriented Approach. New York, NY, USA, Field, Rich and Associates, 1986,47–73.
[90] Fritsche HA, Adams-Park HR. Cellulose acetate electrophoresis of alkaline phosphatase isoenzymes in human serum and tissue. Clin Chem 1972,18,417–21.
[91] Fishman W, Sie HG. L-Homoarginine; an inhibitor of serum "bone and liver" alkaline phosphatase. Clin Chim Acta 1970,29,339–41.
[92] Vinet B, Zizian L, Gauthier B. Characteristics of the inhibition of serum alkaline phosphatase by theophuline. Clin Biochem 1977,11,57–61.
[93] Forsman RW, O'Brien JF. Quantifying bone and liver alkaline phosphatase by the resolution of two-component inactivation data obtained with a centrifugal analyzer. Clin Chem 1991,37, 347–50.
[94] Gomez B Jr, Ardakani S, Ju J, Jenkins D, Cerelli MJ, Daniloff GY, et al. Monoclonal antibody assay for measuring bone-specific alkaline phosphatase activity in serum. Clin Chem 1995,41,1560–6.
[95] Withold W, Schulte U, Reinauer H. Method for determination of bone alkaline phosphatase activity: analytical performance and clinical usefulness in patients with metabolic and malignant bone diseases. Clin Chem 1996,42,210–7.

5 Aspartate aminotransferase and alanine aminotransferase

Joe M. El-Khoury and Sihe Wang

5.1 Case studies

5.1.1 Patient A

A 59-year-old female presented to the headache center with complaints of recurring headaches over the past few months. She had not been sleeping well, her headaches were worsening, and she was experiencing frequent hot flashes. Laboratory studies were unremarkable except for an elevated aspartate aminotransferase (AST) value of 175 U/L. The patient drank alcohol occasionally and exercised frequently. The physician requested that the test be repeated in 2 weeks to confirm the abnormal result and instructed the patient to avoid alcohol use and strenuous exercise during that period. Repeat testing 2 weeks later showed an AST of 164 U/L along with the following results: hepatitis A total antibody, positive; hepatitis A IgM antibody, negative; hepatitis C antibody, negative; iron, 17.4 μmol/L; total iron binding capacity, 48 μmol/L; and transferrin saturation, 36%. The patient was known to have immunity to hepatitis B. The persistently elevated AST was suspected to be caused by certain medications that she was taking. Testing was repeated in another 2 weeks along with a hepatic function panel and showed the following results: activated partial thromboplastin time (APTT), 30.5 s; prothrombin time (PT), 10.8 s; γ-glutamyltransferase (GGT), 15 U/L; albumin, 42 g/L; total bilirubin, 5.1 μmol/L; conjugated bilirubin, 1.7 μmol/L; alkaline phosphatase (ALP), 68 U/L; AST, 197 U/L; alanine aminotransferase (ALT), 23 U/L; total protein, 70 g/L; α_1-antitrypsin, 25.9 μmol/L.

5.1.1.1 Discussion

AST is an enzyme normally found in high activity in the heart, liver, and skeletal muscle. A diseased state in any of these organs is generally accompanied by a variety of clinical symptoms and abnormal biochemical tests. Therefore, after repeating the test to rule out a spurious result due to laboratory error, the patient should be assessed for heart, liver, and skeletal muscle disorders. In the absence of disease in organs rich in AST, the presence of a benign autoimmune condition termed "macro-AST" should be considered. In patients with macro-AST, AST is complexed to immunoglobulins resulting in decreased clearance of AST with a concomitant increase in the half-life and increased AST activity in serum. Isolated persistent increases in AST in asymptomatic patients are highly suggestive of macro-AST. This patient was investigated for the presence of macro-AST by polyethylene glycol precipitation; the result obtained following precipitation was 22 U/L, consistent with the presence of immunoglobulin-bound AST.

5.1.2 Patient B

A 21-year-old White female was brought to the emergency department after being found semi-responsive on the floor of her dormitory room by her roommate. The patient had ingested a full bottle of over-the-counter acetaminophen tablets (~55 g). The time of ingestion was unknown, but was estimated to be no more than 6 h based on the last encounter with her roommate. The acetaminophen concentration 2 h following admission was 1609 µmol/L, and tests in serum revealed the following: AST, 16 U/L; ALT, 30 U/L; total bilirubin, 21 µmol/L; PT, 12 s; creatinine, 274 µmol/L. The patient was started on intravenous N-acetylcysteine (NAC) therapy. After 21 h of NAC therapy, the acetaminophen concentration decreased to 629 µmol/L, whereas serum AST was 30 U/L and ALT was 42 U/L. Acetaminophen concentrations decreased to undetectable levels 78 h after initiation of treatment with NAC, whereas serum aminotransferase activities increased significantly, peaking at around 110 h after initiation of NAC treatment, with values of 4,655 U/L and 6,442 U/L for AST and ALT, respectively. NAC treatment was stopped after aminotransferases returned to baseline values at 220 h post-treatment initiation. She was discharged from the hospital at day 10 after normalization of her laboratory values.

5.1.2.1 Discussion

Hepatotoxicity due to suicidal or therapeutic overdosing of acetaminophen is one of the most common causes of hepatic failure seen in the emergency department. Ingesting massive amounts of acetaminophen may lead to hepatic failure and in some cases acute renal failure, as was observed in this patient by the elevated serum creatinine. Patients with extensive liver damage may require liver transplantation, whereas acute renal failure secondary to acetaminophen-induced hepatic failure is generally reversible. The generally accepted treatment for acetaminophen overdose is administration of NAC for a minimum of 21 h until normalization of acetaminophen, AST, and ALT. Significantly elevated activities (>1,000 U/L) of the aminotransferases is commonly seen in these patients, and this patient required prolonged administration of NAC (220 h) until AST and ALT normalized and acetaminophen was no longer detectable.

5.1.3 Patient C

A 67-year-old White male was admitted to the emergency department after being found stuck between his bed and the wall. The patient lived alone and had been in that state for 2 days before his neighbor heard him screaming and the contacted emergency medical services. The patient described having tripped over a small chair before falling into that position. Relevant medical history included liver cirrhosis secondary to hepatitis C and a history of thrombocytopenia secondary to liver cirrhosis. The patient did not smoke or drink alcohol.

Physical examination revealed mild respiratory distress with jaundiced skin and icteric eyes. His complete blood count (CBC) results showed: white blood cell (WBC) count, 3.5 k/µL; red blood cell (RBC) count, 3.16 M/µL; hemoglobin, 113 g/L; hematocrit 33.5%; platelet count, 44 k/µL; mean cell volume (MCV), 106.0 fL; mean cell hemoglobin (MCH), 35.8 pg; mean cell hemoglobin concentration (MCHC), 337 g/L; red blood cell distribution width (RDW), 17.5%. Coagulation results were: PT, 21.6 s; PT international normalized ratio (INR), 2.1; APTT, 37.7 s. Serum chemistry results showed: AST, 102 U/L; ALT, 41 U/L; ALP, 146 U/L; creatine kinase (CK), 708 U/L; total bilirubin, 210 µmol/L; albumin 21 g/dL; glucose, 5.55 mmol/L.

5.1.3.1 Discussion

Based on the history and laboratory data, particularly the increased AST and CK, a diagnosis of acute rhabdomyolysis was made. The abnormal CK and AST were due to destruction of skeletal muscle fibers. On the following day, AST and CK values decreased to 70 U/L and 508 U/L, respectively, and the AST value dropped to 14 U/L at discharge a week later.

5.1.4 Patient D

A 36-year-old White male was referred to the hepatologist due to abnormal liver enzyme results. The patient was noted to be a heavy smoker (1 pack/day for 20 years) and a heavy drinker (3 drinks/day for 10 years). Physical examination revealed icteric sclera and the patient complained of abdominal pain and nausea.

His CBC results were: WBC, 7.5 k/µL; RBC, 3.85 M/µL; hemoglobin, 133 g/L; hematocrit 37.3%; platelet count, 155 k/µL; MCV, 96.9 fL; MCH, 34.5 pg; MCHC, 357 g/L; RDW, 17.3%. Coagulation results were: PT, 16.2 s; PT INR, 1.42. Serum chemistry results were: AST, 112 U/L; ALT, 39 U/L; ALP, 186 U/L; total bilirubin, 361 µmol/L; albumin, 29 g/L; glucose, 5.0 mmol/L. The patient was negative for hepatitis A, B, and C.

5.1.4.1 Discussion

The patient's history, clinical presentation, and biochemical tests were indicative of alcoholic hepatitis as a result of chronic alcohol ingestion. A minor elevation in AST and an AST/ALT ratio greater than 2.0 are common findings in these patients. The greater increase in AST compared with ALT was due to the release of cytoplasmic and mitochondrial AST from hepatocytes, compared with release of ALT, which is present only in the cytoplasm. Cytoplasmic ALT activities are greater than those of AST, but the combination of AST from both the cytoplasm and mitochondria in alcoholic hepatitis results in much higher AST activities compared with ALT.

The patient was started on prednisolone and was advised to discontinue use of alcohol and to quit smoking. After remaining sober for 3 months, the patient's AST and ALT values reverted back to normal (33 U/L and 28 U/L, respectively).

5.1.5 Patient E

A 26-year-old White male presented to the hospital with a sore throat, nausea, vomiting, fatigue, and weight loss, which started approximately 2 weeks ago. The patient had icteric sclera and an enlarged liver as seen in his physical and sonographic examination. The patient used alcohol rarely and was noted to be a tattoo artist.

Serum chemistry results were: AST, 325 U/L; ALT, 437 U/L; total bilirubin, 50 µmol/L; direct bilirubin, 15 µmol/L. The patient tested negative for hepatitis A and B antibodies, but was positive for both hepatitis C antibody and HCV-RNA.

The patient was started on interferons and 2 months later his aminotransferase activities in serum were: AST, 28 U/L and ALT, 23 U/L. After 10 months of treatment, the patient tested negative for HCV-RNA and his aminotransferases were still in the normal ranges.

5.1.5.1 Discussion

This patient had chronic hepatitis C. The most likely explanation of the infection was that the patient contracted the virus while working on a tattoo for one of his customers who was positive for hepatitis C. There are several reported cases in the literature of healthcare workers contracting hepatitis C, mainly through needlestick accidents and blood splashes to the eyes. The abnormal aminotransferases indicate damage to the hepatocytes, and bilirubin was increased to where icteric sclera was visible. A hepatitis panel is usually indicated when liver dysfunction is suspected without an apparently known cause. After treatment with interferons, the aminotransferases return to normal within a few weeks to a month.

5.2 Biochemistry and physiology

5.2.1 Molecular forms

AST (EC 2.6.1.1), a vitamin B6-dependent enzyme, belongs to the class I pyridoxal phosphate-dependent aminotransferase family. It is virtually ubiquitous in eukaryotic cells and is primarily located in the heart, liver, skeletal muscle, and kidneys with lesser activities in other organs. AST, also known as glutamate oxaloacetate transaminase (GOT), exists in two distinct forms: cytoplasmic (c-AST or GOT1) and mitochondrial (m-AST or GOT2). The amino acid sequences and structures of these isoenzymes have been extensively studied in various species including humans [1]. Both isoenzymes possess a homodimeric structure composed of two identical polypeptide subunits that are coded for by distinct but structurally related genes [2]. c-AST consists of 413 amino acids per polypeptide [3], whereas m-AST consists of 430 amino acids per polypeptide [4]. Similar to most nuclear-encoded mitochondrial proteins, m-AST is synthesized as a larger precursor (pre-m-AST) in the cytosol first, then translocated and cleaved in the mitochondrial membrane to form the mature m-AST, whereas c-AST has no known precursor [2]. Three to five subforms of c-AST have been identified by isoelectric focusing, while only one form of m-AST has been found [5, 6]. The molecular weight of the holoenzymes isolated from human liver was determined to be 93,000 for c-AST and 90,400 for m-AST [7]. Other biochemical and structural differences of the two enzymes are summarized in Table 5.1.

The association of enzymes with immunoglobulins (mainly IgA or IgG), a phenomenon called macroenzyme formation, leads to an increased molecular weight and a decreased clearance rate of the enzyme, resulting in persistently elevated enzyme levels [8]. Macroenzymes have been associated with several diseases, mainly autoimmune disorders, but there is no convincing evidence that antienzyme antibodies cause the diseases [9]. AST complexed with immunoglobulins, macro-AST, is generally considered a non-pathological condition. However, persistently elevated AST activities may result in unnecessary costly and invasive diagnostic procedures in order to determine the underlying cause [10]. Macro-AST was first reported in 1978 [11] and has since been increasingly identified in children [12–14] and adults [10, 15, 16], generally with an isolated elevation of AST activity without any clinical presentation. There are several methods for detecting macro-AST, including electrophoresis, immunoprecipitation with polyethylene glycol, measurement of heat stability, gel filtration chromatography, and measuring recovery of

Table 5.1: Biochemistry of cytoplasmic and mitochondrial AST.

	Cytoplasmic	Mitochondrial
Number of subforms [5, 6]	3–5	1
Heat stability	Poor	Good
Inhibition by phosphate	No	Yes
Serum half-life activity of enzyme after an acute myocardial infarction [17]	20 h	35 h
Electrophoretic mobility	Anodal	Cathodal
Isoelectric point [6, 18]	5.15–5.8	>9.5

enzyme activity after IgG or IgA binding with protein G or protein A beads [10, 15]. Identification by electrophoresis is considered the gold standard [13] due to the unique migrating pattern of macro-AST, whereas identification by repeat measurement of enzyme activity after protein A and protein G absorption provides a simple and convenient alternative. It is highly recommended that clinicians consider macro-AST in the diagnosis of an asymptomatic patient with an isolated elevation in AST activity, and carefully document this condition in the patient's records to avoid future time-consuming, expensive, and/or invasive procedures [10, 13, 19, 20]. In addition, it is important to recognize macro-AST as a benign condition and reassure the patient that no specific treatment is necessary [21].

Similar to AST, ALT (EC 2.6.1.2) is also a vitamin B6 dependent enzyme belonging to the pyridoxal phosphate-dependent aminotransferase family. However, unlike AST, ALT exhibits relatively much higher activity in the liver compared to other tissues (see Table 5.2), followed by activity in the kidneys and heart. In addition, ALT exists in one molecular form only, which is located in the cytoplasm and consists of a sequence of 495 amino acids [22].

5.2.2 Biochemical function

AST catalyzes the transfer of an amino group between L-aspartate and 2-oxoglutarate to form oxaloacetate and L-glutamate in a reversible reaction. The enzyme requires P-5'-P as a cofactor, which accepts the amino group from L-aspartate forming pyridoxamine-5'-phosphate and oxaloacetate. Next, pyridoxamine-5'-phosphate transfers the amino group to 2-oxoglutarate to form glutamate. The reaction can be summarized as follows, Equation (5.1):

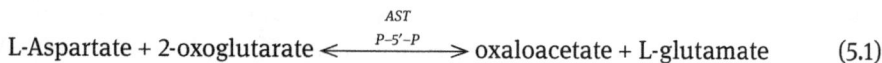

$$\text{L-Aspartate} + \text{2-oxoglutarate} \xrightleftharpoons[P\text{-}5'\text{-}P]{AST} \text{oxaloacetate} + \text{L-glutamate} \tag{5.1}$$

The mechanism of catalysis of AST based on structural and kinetic studies are well described in the literature [23–26]. Briefly, the reaction starts with the formation of an aldimine intermediate between the amino acid and P-5'-P, then converts to a quinoid structure which breaks down to a ketimine and later decomposes to pyridoxamine and the keto acid. In the reverse reaction, the pyridoxamine reacts with ketoacid to give P-5'-P and the amino acid.

ALT catalyzes the reaction between L-alanine and 2-oxoglutarate and uses P-5'-P as a cofactor in the same way AST does, and its reaction is illustrated as the following, Equation 5.2:

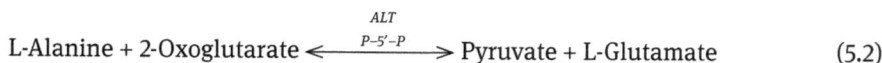

$$\text{L-Alanine} + \text{2-Oxoglutarate} \xrightleftharpoons[P\text{-}5'\text{-}P]{ALT} \text{Pyruvate} + \text{L-Glutamate} \tag{5.2}$$

Table 5.2: Tissue activities of AST and ALT relative to serum.

Tissue	AST relative activity	ALT relative activity
Heart	7,800	444
Liver	7,100	2,750
Skeletal muscle	4,950	300
Kidney	4,550	1,188
Pancreas	1,400	125
Spleen	700	75
Lung	500	44
Erythrocytes	8	4
Normal serum	1	1

Although many transaminases have been described, only ALT and AST are found to have serum activities that correlate with disease processes in organs like liver, skeletal muscle and heart.

5.2.3 Normal physiology

As can be inferred from the activity values shown in Table 5.2, AST is predominantly located in the heart, liver, skeletal muscle, and kidneys. In serum the activity of m-AST is very low (<10%), whereas in the liver m-AST accounted for 80% of total enzyme activity [27, 28]. On the other hand, ALT is predominantly located in the liver with much lesser activities in kidneys and heart and is exclusively located in the cytoplasm.

Both the holoenzyme and apoenzyme of AST and ALT are present in blood. Deficiency in vitamin B6 can result in reduction of apparent serum AST and ALT activities owing to the decreased availability of their cofactor pyridoxal-5'-phosphate P-5'-P. However, this is only true for assays that do not incorporate P-5'-P within the assay reagent.

5.2.3.1 Elimination

The mechanism of AST clearance has not been well studied; however, experiments in rats have demonstrated that the liver plays a central role in the process and *in vitro* studies have shown that uptake occurs exclusively by sinusoidal liver cells with a much faster clearance rate for m-AST than for c-AST [29]. The half-life for AST in circulation is 12–22 h, while ALT is cleared more slowly and has a much longer half-life of 37–57 h [30].

5.2.3.2 Exercise

The extent of increase of serum enzyme activities following exercise depends on the strength and duration of the exercise, the physical condition of the person, and time of blood collection. Strenuous exercise may increase serum AST as much as three-fold, with the effect more prominent in males [30]. On the other hand, moderate exercise decreases ALT by 20% in comparison with individuals who do not exercise or those who exercise strenuously [30]. Individuals who train regularly, such as athletes, have a lesser increase in enzyme levels when compared with untrained individuals.

5.2.4 Reference ranges

The reference range for AST and ALT was first determined by Siest et al. in 1975 [31]. However, these ranges are no longer valid due to the changes implemented to the measurement procedure by the International Federation of Clinical Chemistry (IFCC) in an attempt for standardization of AST and ALT measurements. The most recent change to the reference procedure was a redefinition of the reference temperature from 30°C to 37°C [32]. To expedite the transition to the new IFCC reference procedure, a preliminary upper limit of reference range (ULR) was determined using 837 hospitalized individuals (419 females and 418 males) older than 17 years [33]. The AST ULR (97.5th percentile) for females and males was determined to be 31 U/L and 35 U/L, respectively. The ALT ULR for females and males was determined to be 34 U/L and 45 U/L, respectively. However, a later investigation involving 765 individuals (411 females and 354 males) in four different regions worldwide, selected based on results of other laboratory tests and a specific questionnaire, found that the gender difference of ULRs was only 1.7 U/L, making it possible to have a single reference interval (11–34 U/L) [34]. However, ALT still showed significant gender difference with the reference intervals determined to be 8–41 U/L for females and 9–59 U/L for males. Interregional differences for both enzymes were found statistically insignificant.

Kim et al. [35] conducted a prospective cohort study and found a positive association between the ULR of AST and mortality from liver disease in a population of 142,055 individuals aged between 35 and 59 years, selected from a Korean health insurance corporation and followed from 1993 to 2000. The adjusted relative risks for AST concentration of 20–29 U/L and 30–39 U/L compared with AST levels <20 U/L were 2.5 and 8.0 in males and 3.3 and 18.2 in females, respectively. The authors estimated that the best cut-off value for identifying risk of death from liver disease for males is 31 U/L, whereas no cut-off value was established for females because few died from liver disease in this study population. However, the procedures used for measuring AST most likely do not conform to the new IFCC reference procedures published in 2002.

5.3 Chemical pathology

Measurement of ALT and AST is primarily used for the detection of acute or chronic liver injury. Other important causes of AST increases include skeletal muscle and cardiac injury, but the availability of more sensitive and specific biomarkers makes the measurement of AST less useful and cost-ineffective in these clinical situations. ALT is more specific to liver injury because of its high abundance relative to other organs and is therefore more useful than AST as a marker for liver cell injury. Other less common causes of isolated AST and ALT elevation should also be investigated in the absence of injury to the main organs that are rich in AST and ALT.

5.3.1 Liver disease

Liver disease is one of the most important and common causes of abnormal enzyme activities in serum. Liver biochemical abnormalities can be categorized into three categories: (i) hepatocellular damage, (ii) cholestasis, or (iii) impaired synthetic or metabolic function. Increased levels of the aminotransferases, ALT and AST, indicate hepatocyte injury and necrosis. ALT is more specific than AST for liver damage, and in most types of liver diseases serum ALT is higher than AST except in alcoholic hepatitis, hepatic cirrhosis, and liver cancer [36]. A more detailed classification is described in Table 5.3 [37–39]. Although helpful, it should be noted that the absolute increase in serum aminotransferase activity is not specific for the cause of liver disease, and many diseases may span across all three levels of

Table 5.3: Common causes of elevated serum aminotransferases[1].

Elevation	Cause
Minor (<100 IU/L)	Chronic hepatitis C
	Chronic hepatitis B
	Hemochromatosis
	Fatty liver
	Celiac sprue
Moderate (100–300 IU/L)	As above plus
	Alcoholic hepatitis
	Non-alcoholic steatohepatitis
	Autoimmune hepatitis
	Wilson's disease
Major (>1,000 IU/L)	Autoimmune hepatitis
	Acute viral hepatitis
	Ischemic or toxic injury

Note: 1. Reproduced, with permission, from [37]. Copyright © 2002 Royal College of Physicians.

transaminase elevation (Table 5.3) [39]. The duration of enzyme abnormalities, the peak values, the changes over time, and the values in the context of the clinical state are important. Declining aminotransferase values in the face of an improving or deteriorating clinical picture have very different meanings. In the former it generally implies recovery, in the latter it suggests widespread necrosis, hepatic failure, and a poor prognosis.

5.3.1.1 Ischemic and toxic liver injury

The most common causes of significantly elevated AST and ALT activities are either ischemic or toxic liver injury, especially drugs and toxins such as acetaminophen. Acetaminophen poisoning accounts for almost one-half of all the cases of acute liver failure in the USA and the UK [40]. In 90% of acetaminophen-induced liver injury cases, AST and ALT are greater than 3,000 U/L, and typically peaks within the first 24 h after admission before declining rapidly and reaching near normal activities 7 days following onset of injury [41, 42]. Many drugs may cause significant increases in AST and ALT. In some cases, hepatic injury may not become evident until as much as 12 months following initiation of treatment, which is why it is important to ask patients about all drugs received within the past year [43]. An easy way to determine if the drug is causing an increase in AST or ALT is to request the patient to stop taking the medication and see if aminotransferase levels return to normal. However, this requires that the physician make a risk-benefit analysis to determine whether the drug should be continued despite the elevation in aminotransferases [44]. Common drugs falling into this category include non-steroidal anti-inflammatory drugs, antibiotics, anti-epileptic drugs, anti-tuberculosis drugs, and inhibitors of hydroxymethylglutaryl-coenzyme A reductase [44]. Drugs and herbal preparations that have been shown to cause elevations in liver enzyme levels have been summarized in a recent report [44].

5.3.1.2 Viral hepatitis

In acute viral hepatitis, AST and ALT activities peak before the appearance of clinical symptoms such as jaundice, but decrease much more gradually to normal levels, usually within 3–5 weeks [36]. Generally, hepatitis A and B show a greater increase in aminotransferase levels compared with hepatitis C; the former are self-limited illnesses that resolve completely in most patients, whereas the latter has an 85% chance of developing into a chronic condition [38, 43]. It is highly recommended that the initial evaluation of a patient with acute hepatic injury includes testing for antibodies to hepatitis A, B, and C virus, followed by hepatitis D in patients with presence of hepatitis B surface antigen [38, 43]. Other less common causes include herpesvirus, cytomegalovirus, enterovirus, coronavirus, reovirus (in neonates), adenovirus, parvovirus B6 (in children), varicella-zoster virus, and Epstein-Barr [43].

Wai et al. [45] developed a tool to enable prediction of fibrosis and cirrhosis in chronic hepatitis C patients. The tool uses two routinely measured laboratory tests to calculate the AST to platelet ratio index (APRI), Equation (5.3).

$$APRI = \frac{AST\ level\ (/ULR)}{Platelet\ counts\ (10^9/L)} \times 100 \qquad (5.3)$$

Using a scale from 0 to 10, scores less than 0.5 had a negative predictive value of 86%, whereas scores greater than 1.5 had a positive predictive value of 88% for identifying significant fibrosis [46].

5.3.1.3 Autoimmune hepatitis

Autoimmune hepatitis (AIH) is a generally unresolved inflammation of the liver of unknown cause. The disease is highly prevalent in young and middle-aged women with a female to male ratio of 3.6:1 [44, 47]. Diagnostic criteria for AIH were simplified in 2008 by the International Autoimmune Hepatitis Group [48]. The previously used scoring system from 1999 was cumbersome because it categorized individuals into three groups (not AIH, probable AIH, and definite AIH) based on 12 variables, including the AST : ALT ratio [49]. The new and simplified scoring system uses only four variables and does not include the AST : ALT ratio. Studies comparing the performance of the simplified 2008 scoring system to the 1999 scoring system found that the latter had better sensitivity (93% vs. 100%, respectively), whereas the former had better specificity (98% vs. 97%, respectively) [50, 51]. Other useful indicators are hypergammaglobulinemia, which is present in as many as 80% of AIH patients, autoantibodies, viral hepatitis (seronegative for hepatitis A, B, and C) and liver biopsy [47].

5.3.1.4 Alcoholic hepatitis

Identification of alcoholic hepatitis can be challenging due to the difficulty in obtaining an accurate drinking history from the patient; therefore, laboratory tests that can help establish this diagnosis are highly desirable. The most frequently used markers of alcohol consumption are GGT, AST and ALT, and the erythrocyte MCV, with the serum GGT being the most sensitive and most widely used [52]. However, GGT elevation is also associated with non-alcohol-related diseases and is not a specific biomarker when used by itself.

Other useful aids in determining the cause of hepatic injury are the ratios of AST/ALT, m-AST/AST, and the more recently introduced APRI. In a study by Cohen and Kaplan [53], the AST/ALT ratio was greater than 1 in 92% and greater than 2 in 70% of patients with alcoholic liver disease. This observation may be explained by two mechanisms: (i) patients with alcoholic dependence are often deficient in P-5'-P, which affects ALT more than AST in methods that do not include P-5'-P as a reagent in the assay and (ii) ethanol upregulates plasma membrane expression and export of m-AST, as demonstrated in HepG2 cells [54]. In healthy individuals, m-AST/AST is typically less than 0.1, whereas in alcoholic liver disease this ratio increases significantly. In conjunction with an AST/ALT ratio greater than 2, the m-AST/AST ratio provides further evidence of alcoholic liver disease [55]. In addition, the APRI may also be used based on the observation that AST is elevated whereas platelet counts are reduced in patients with alcoholic hepatitis.

5.3.1.5 Non-alcoholic fatty liver disease
Non-alcoholic fatty liver disease (NAFLD) is the most common cause of hepatic injury other than viral and alcoholic hepatitis. It occurs most frequently in obese middle-aged women and is histologically categorized into four types: (i) type 1, fat alone; (ii) type 2, fat plus inflammation; (iii) type 3, fat plus ballooning degeneration; and (iv) type 4, fat plus fibrosis and/or Mallory bodies [56]. Liver biopsy is considered the gold standard for diagnosis, but other non-invasive tests that correlate well with the degree of fibrosis and cirrhosis are highly desired [57]. An AST/ALT ratio of greater than 0.8, in combination with elevated ALT, blood pressure, triglycerides, and insulin resistance index is considered a strong predictor of fibrosis [56]. This ratio is reported to be less than 1 in 65% to 90% of patients with NAFLD, and a value >1 suggests an advanced fibrotic form of NAFLD, whereas it is almost never greater than 2 in this subset of patients [56]. In a study involving 70 patients with non-alcoholic steatohepatitis (NASH) and 70 patients with alcoholic liver disease, the AST/ALT ratio was useful in distinguishing the two populations, with a mean AST/ALT of 0.9 in patients with NASH and 2.6 in patients with alcoholic liver disease [58].

5.3.2 Hemochromatosis

Hereditary hemochromatosis (HH) is a common autosomal recessive disorder that is characterized by toxic iron deposition in the liver, pancreas, and heart. HH can be a result of mutations in several genes, including *HFE, TfR2, HJV, HAMP,* and *Ferroportin* [59]. Currently, the majority of HH cases are diagnosed after the evaluation of ferritin concentrations and the serum transferrin-iron saturation index, which if greater than 45% is highly suggestive of this disease [34, 48, 59]. Although the need for liver biopsy has decreased with the availability of genetic testing, biopsy remains the main diagnostic tool due to non-HFE-related hemochromatosis [38]. However, non-invasive markers for detecting fibrosis and cirrhosis are being actively investigated in patients with HH. In a retrospective study involving 32 HH patients, platelet counts of 200×10^9 or less and an elevated AST revealed a negative predictive value of 100% for high-degree fibrosis [61]. In another study involving a cohort of 193 Canadian and 162 French C282Y homozygous patients, the combination of ferritin greater than or equal to 1000 µg/L, a platelet count of 200×10^9 or less, and an elevated AST was useful in diagnosing an average of 81% of cases with cirrhosis [61]. Liver fibrosis may also be present in 18% of C282Y homozygotes with normal AST and ALT [62].

5.3.2.1 Wilson's disease

Wilson's disease is a genetic disorder of biliary copper excretion, usually detected between the ages of 5 and 25 years, but also considered in patients up to 40 years of age [44]. Clinically, patients should be examined for Kayser-Fleischer rings, and biochemically serum ceruloplasmin will be reduced in 85% of affected patients [44, 63]. The disease can also cause the AST/ALT ratio to exceed 4 [64]. However, a study involving patients with acute liver failure due to Wilson's disease (n = 7) and those due to other etiologies (n = 8) found that an AST/ALT ratio greater than 4 showed no utility for differentiating between the two groups, whereas both AST and ALT activities were lower in patients with Wilson's disease [65]. In a cohort of 140 acute liver failure patients, an AST/ALT greater than 2.2 yielded a sensitivity of 94% and a specificity of 86% for predicting Wilson's disease [66]. The authors also showed that combining the more readily available tests such as ALP, bilirubin, and the aminotransferases gave a diagnostic sensitivity and specificity of 100% for identifying Wilson's disease in acute liver failure patients.

5.3.2.2 α$_1$-Antitrypsin deficiency

α$_1$-Antitrypsin (A1AT) deficiency occurs in approximately 1 per 1,000 to 1 per 2,000 persons of European ancestry and is an uncommon cause of chronic liver disease in adults [43]. It is an uncommon cause of AST elevation and can be diagnosed by either direct measurement of serum levels of A1AT or by the lack of an α$_1$-globulin band in serum protein electrophoresis [44].

5.3.3 Skeletal muscle disease

AST is present in high activity in skeletal muscle. In disorders affecting skeletal muscle, AST is elevated along with CK, lactate dehydrogenase, aldolase, myoglobin, and carbonic anhydrase III [67].

AST is also increased in the setting of muscular dystrophy. For patients with unexplained increases in AST, measurement of serum CK and careful physical examination is recommended to identify patients with muscular dystrophies [68].

5.3.3.1 Rhabdomyolysis

Rhabdomyolysis, or destruction of skeletal muscle tissue, is nearly always accompanied by increases in CK, lactate dehydrogenase, AST, and myoglobin. Important causes are trauma, alcohol, seizure disorders, and physical exertion. In a retrospective chart review of 215 cases of rhabdomyolysis with CK ≥1,000 U/L, Weibrecht et al. [69] observed abnormal AST (defined as AST >40 U/L) in 93.1% of cases. Furthermore, AST fell in parallel with CK during the first 6 days of hospitalization for patients with rhabdomyolysis. AST may also be elevated during or after moderate exercise for a prolonged amount of time due to exertional rhabdomyolysis. In an ultra-endurance running exercise, 39 runners who ran a 246-km continuous race had asymptomatic exertional rhabdomyolysis with AST up to 50 times the ULR [70].

5.3.3.2 Myositis

CK is usually measured in cases of polymyositis (PM) and dermatomyositis (DM). However, the literature suggests that in 5–10% of cases of DM and PM, CK can be normal [71]. As a result, it is recommended that aldolase and AST be also measured during the evaluation of a patient with myositis because it is unlikely that all three will be normal [72]. Louthrenoo et al. [73] studied the correlation between muscle enzymes and DM/PM in 100 Thai patients and found that CK, lactate dehydrogenase, and AST were elevated in 87%, 92%, and 82% of cases, respectively.

5.3.4 Heart disease

5.3.4.1 Acute myocardial infarction

The link between AST and acute myocardial infarction (AMI) was first established in 1954 after elevated AST activities were observed in patients with AMI [74]. However, use of AST is no longer recommended for the evaluation of cardiac injury and detection of myocardial infarction. The current guidelines by the American Heart Association and the National Academy of Clinical Biochemistry specify that troponins should be primarily used as the biomarkers of AMI [75, 76].

Both c-AST and m-AST are increased in patients with AMI [8]. After the appearance of clinical symptoms, such as chest pain, c-AST and m-AST peak at 28 h and 52 h, respectively, and peak c-AST but not m-AST correlate with peak CK-MB values, indicating that c-AST also reflects the extent of myocardial tissue damage [77, 78]. Annoni et al. [79] demonstrated that serum m-AST and the ratio of m-AST/AST correctly classified 91.9% of 112 AMI patients as survivors or non-survivors.

5.3.4.2 Other heart diseases

Serum AST is usually elevated in patients with myocarditis and pericarditis. In a retrospective study of a pediatric population of 31 patients under the age of 18 years admitted to the emergency department and diagnosed with myocarditis, AST had the best diagnostic sensitivity (85%) among the laboratory tests measured [80]. Nanji [81] observed significantly increased AST (>8,000 U/L) in two patients with congestive heart failure. With improvement in the patients' circulatory status due to aggressive treatment, AST levels decreased markedly.

5.3.5 Other causes of AST increase

Increased AST in the absence of liver, heart, or skeletal muscle injury should prompt an investigation of less common causes of AST elevation, such as celiac disease. Celiac disease is one of the most common inflammatory disorders of the small intestine with a prevalence of 1% in the Western world [82]. In a study involving 140 consecutive patients with increased aminotransferase activities of unknown cause, 13 were found to have celiac disease after testing positive for anti-gliadin and anti-endomysium antibodies [83].

5.4 Analysis

5.4.1 Specimens

Serum is considered to be the specimen of choice, but heparinized or EDTA plasma may be used for the determination of catalytic activity of AST and ALT.

5.4.2 Reaction used

For AST, the predominantly used reaction scheme is, Equation (5.4):

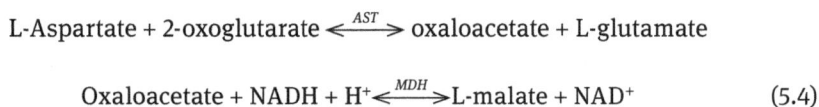

$$\text{L-Aspartate + 2-oxoglutarate} \xleftrightarrow{AST} \text{oxaloacetate + L-glutamate}$$

$$\text{Oxaloacetate + NADH + H}^+ \xleftrightarrow{MDH} \text{L-malate + NAD}^+ \tag{5.4}$$

For ALT, the predominantly used reaction scheme is, Equation 5.5:

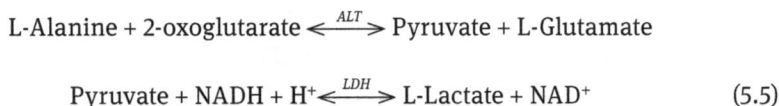

$$\text{L-Alanine + 2-oxoglutarate} \xleftrightarrow{ALT} \text{Pyruvate + L-Glutamate}$$

$$\text{Pyruvate + NADH + H}^+ \xleftrightarrow{LDH} \text{L-Lactate + NAD}^+ \tag{5.5}$$

The reaction rate is monitored at 340 nm, which corresponds to a change in NADH. As a result, the rate of NADH consumed, as measured spectrophotometrically, is directly proportional to AST and ALT activity. AST and ALT are stable for 24 h at 15°C to 35°C [84] and 7 days at 2°C to 8°C [2], and prolonged exposure to blood cells up to 56 h at 25°C had no significant impact [85].

5.4.2.1 Primary reference measurement procedure

For more than two decades, the IFCC has tried to increase the widespread use of its recommended procedures for the measurement of AST and ALT [86–89]. It then became clear that the goal of a single, universally accepted procedure for the measurement of the catalytic activity of an enzyme will never be achieved, and establishing a reference system in clinical enzymology would be a more feasible alternative to reduce inter-laboratory variation [90]. As a result, the IFCC primary reference measurement procedures (PRMPs) for the international standardization of catalytic concentration measurements of several enzymes including AST and ALT have been published [32, 91–96]. These publications contain detailed descriptions of the procedure, which have been more recently updated [97]. The main change from the original IFCC procedure published in 1986 was as change in measurement temperature from 30°C to 37°C.

5.4.2.2 Certified reference material

Although the IFCC PRMPs for AST and ALT were published, lower metrological levels, such as those in routine use, could not be traced to the PRMPs because of the unavailability of a certified reference material (CRM). Recently, the IFCC and the Institute for Reference Materials Measurements developed a CRM for AST and ALT (ERM-AD457/IFCC), which is lyophilized from a mixture of a human type recombinant enzyme expressed in *Escherichia coli* and a buffer containing bovine serum albumin [98]. Creating this material required high quality processing technology and it has been extensively evaluated in terms of stability and homogeneity by international laboratories. In combination with PRMPs, the CRM will allow the standardization of AST and ALT measurements, thus improving accuracy and reducing inter-laboratory variation.

5.4.3 Interferences

Samples displaying mild or almost undetectable hemolysis by visual inspection can exhibit clinically meaningful changes in AST [99]. Hence, samples with visible hemolysis should prompt the laboratory to alert the physician and to recollect the sample.

Phosphate buffers should be avoided because they retard the reaction of apo-AST and P-5′-P [19]. Reagents containing ammonium salts must be avoided due to its impact on the reaction catalyzed by glutamate dehydrogenase in serum to form glutamate and consume NADH, which may lead to falsely elevated measured AST activity [19]. Endogenous serum ammonia has a negligible effect on AST activity, even when abnormal [19].

Hydroxocobalamin given intravenously at therapeutic levels for the treatment of cyanide poisoning significantly interferes with the analytical measurement of AST [100, 101].

5.4.4 Methods for AST isoenzymes

The methods for determining AST isoenzymes in human serum have been extensively reviewed by Panteghini [2]. In general, the methods described fall under four categories: differential chemical inhibition, electrophoresis, chromatography, and immunoassays. Immunochemical techniques were found to be the most accurate and precise for the measurement of AST in human serum, but electrophoresis methods are useful for the detection of atypical forms (genetic variants, macroenzyme complexes) of AST [18].

5.5 Questions and answers

1. AST requires what as a cofactor?
   (a) Magnesium
   (b) Vitamin B3
   (c) Vitamin B6
   (d) Calcium
   (e) Vitamin B12
2. Where is AST located intracellularly?
   (a) Cytoplasm
   (b) Nucleus
   (c) Mitochondria
   (d) Cytoplasm and mitochondria
   (e) Cytoplasm and nucleus
3. Which statement is false about c-AST and m-AST?
   (a) Both c-AST and m-AST have a homodimeric structure composed of two identical polypeptide subunits
   (b) m-AST is synthesized as a precursor first, whereas c-AST has no known precursor
   (c) There are multiple subforms of c-AST identified, but only one form of m-AST
   (d) The holoenzyme of m-AST has a larger molecular weight than c-AST
   (e) m-AST has a longer serum activity half-life after myocardial infarction than c-AST
4. A persistently elevated level of AST in asymptomatic patients is suggestive of what condition?
   (a) Myocardial infarction
   (b) Macro-AST
   (c) Ischemic or toxic liver injury
   (d) Rhabdomyolysis
   (e) Renal failure

5. In what tissues is AST predominantly located?
   (a) Heart, lungs, skeletal muscle, and kidneys
   (b) Liver, lungs, skeletal muscle, and kidneys
   (c) Heart, liver, duodenum, and kidneys
   (d) Liver, duodenum, skeletal muscle, and kidneys
   (e) Heart, liver, skeletal muscle, and kidneys

6. What is the most common cause for elevated AST?
   (a) Ischemic or toxic liver injury
   (b) Myocardial infarction
   (c) Rhabdomyolysis
   (d) Viral hepatitis
   (e) Autoimmune hepatitis

7. How is AST measured?
   (a) By spectrophotometrically monitoring the rate of disappearance of NAD^+
   (b) By spectrophotometrically monitoring the rate of appearance of malate
   (c) By spectrophotometrically monitoring the rate of disappearance of NADH
   (d) By spectrophotometrically monitoring the rate of disappearance of oxaloacetate
   (e) By spectrophotometrically monitoring the rate of appearance of NAD^+

8. How did the IFCC attempt to improve inter-laboratory variation for AST?
   (a) Establish recommended procedures
   (b) Establish a primary reference measurement procedure
   (c) Publish procedural notes and useful advice
   (d) Develop certified reference materials
   (e) All of the above

9. Which of the following statement is correct for ALT?
   (a) ALT exists in both cytoplasm and mitochondria
   (b) ALT display higher activity in liver compared to AST
   (c) ALT is more specific than AST as a biomarker for liver cell injury
   (d) In hepatitis patients plasma AST level is always higher than ALT
   (e) ALT activity is the highest in kidney

References

[1] Schneider G, Kack H, Lindqvist Y. The manifold of vitamin B6 dependent enzymes. Structure 2000,8,R1–6.
[2] Panteghini M. Aspartate aminotransferase isoenzymes. Clin Biochem 1990,23,311–9.
[3] Doyle JM, Schinina ME, Bossa F, Doonan S. The amino acid sequence of cytosolic aspartate aminotransferase from human liver. Biochem J 1990,270,651–7.
[4] Venter JC, Adams MD, Myers EW, Li PW, Mural RJ, Sutton GG, et al. The sequence of the human genome. Science 2001,291,1304–51.

[5] Leung FY, Henderson AR. Multiple molecular forms of human heart cytoplasmic aspartate aminotransferase. Clin Sci (Lond) 1982,62,337–9.

[6] Rej R. Multiple molecular forms of human cytoplasmic aspartate aminotransferase. Clin Chim Acta 1981,112,1–11.

[7] Rej R. Measurement of aminotransferases. Part 1. Aspartate aminotransferase. Crit Rev Clin Lab Sci 1984,21,99–186.

[8] Galasso PJ, Litin SC, O'Brien JF. The macroenzymes: a clinical review. Mayo Clin Proc 1993,68,349–54.

[9] Remaley AT, Wilding P. Macroenzymes: biochemical characterization, clinical significance, and laboratory detection. Clin Chem 1989,35,2261–70.

[10] Krishnamurthy S, Korenblat KM, Scott MG. Persistent increase in aspartate aminotransferase in an asymptomatic patient. Clin Chem 2009,55,1573–5.

[11] Konttinen A, Murros J, Ojala K, Salaspuro M, Somer H, Rasanen J. A new cause of increased serum aspartate aminotransferase activity. Clin Chim Acta 1978,84,145–7.

[12] Wiltshire EJ, Crooke M, Grimwood K. Macro-AST: a benign cause of persistently elevated aspartate aminotransferase. J Paediatr Child Health 2004,40,642–3.

[13] Caropreso M, Fortunato G, Lenta S, Palmieri D, Esposito M, Vitale DF, et al. Prevalence and long-term course of macro-aspartate aminotransferase in children. J Pediatr 2009,154,744–8.

[14] Fortunato G, Iorio R, Esposito P, Lofrano MM, Vegnente A, Vajro P. Macroenzyme investigation and monitoring in children with persistent increase of aspartate aminotransferase of unexplained origin. J Pediatr 1998,133,286–9.

[15] Stasia MJ, Surla A, Renversez JC, Pene F, Morel-Femelez A, Morel F. Aspartate aminotransferase macroenzyme complex in serum identified and characterized. Clin Chem 1994,40,1340–3.

[16] Werner T, Vargas HE, Chalasani N. Macro-aspartate aminotransferase and monoclonal gammopathy: a review of two cases. Dig Dis Sci 2007,52,1197–8.

[17] Niblock AE, Jablonsky G, Leung FY, Henderson AR. Changes in mass and catalytic activity concentrations of aspartate aminotransferase isoenzymes in serum after a myocardial infarction. Clin Chem 1986,32,496–500.

[18] Leung FY, Henderson AR. Isolation and purification of aspartate aminotransferase isoenzymes from human liver by chromatography and isoelectric focusing. Clin Chem 1981,27,232–8.

[19] Lott JA, Wolf PL. Alanine and aspartate aminotransferase (ALT and AST). In: Lott JA, Wolf PL, eds. Clinical Enzymology: a Case-oriented Approach. New York, NY, USA, Field, Rich, and Associates, 1986,111–38.

[20] Mari T, Morini S, Zullo A, Marignani M. Macroenzyme: do not forget liver function tests. Dig Liver Dis 2003,35,673.

[21] Sass DA, Chadalavada R, Virji MA. Unexplained isolated elevation in serum aspartate aminotransferase: think macro! Am J Med 2007,120,e5–6.

[22] Ishiguro M, Takio K, Suzuki M, Oyama R, Matsuzawa T, Titani K. Complete amino acid sequence of human liver cytosolic alanine aminotransferase (GPT) determined by a combination of conventional and mass spectral methods. Biochemistry 1991,30,10451–7.

[23] Hammes GG. Multiple conformational changes in enzyme catalysis. Biochemistry 2002,41,8221–8.

[24] Kirsch JF, Eichele G, Ford GC, Vincent MG, Jansonius JN, Gehring H, et al. Mechanism of action of aspartate aminotransferase proposed on the basis of its spatial structure. J Mol Biol 1984,174,497–525.

[25] Jansonius JN, Eichele G, Ford GC, Kirsch JF, Picot D, Thaller C, et al. Crystallographic studies on the mechanism of action of mitochondrial aspartate aminotransferase. Prog Clin Biol Res 1984,144B,195–203.

[26] Picot D, Sandmeier E, Thaller C, Vincent MG, Christen P, Jansonius JN. The open/closed conformational equilibrium of aspartate aminotransferase. Studies in the crystalline state and with a fluorescent probe in solution. Eur J Biochem 1991,196,329–41.

[27] Rej R. Aspartate aminotransferase activity and isoenzyme proportions in human liver tissues. Clin Chem 1978,24,1971–9.

[28] Rej R. Aminotransferases in disease. Clin Lab Med 1989,9,667–87.

[29] Kamimoto Y, Horiuchi S, Tanase S, Morino Y. Plasma clearance of intravenously injected aspartate aminotransferase isozymes: evidence for preferential uptake by sinusoidal liver cells. Hepatology 1985,5,367–75.

[30] Dufour DR, Lott JA, Nolte FS, Gretch DR, Koff RS, Seeff LB. Diagnosis and monitoring of hepatic injury. I. Performance characteristics of laboratory tests. Clin Chem 2000,46,2027–49.

[31] Siest G, Schiele F, Galteau MM, Panek E, Steinmetz J, Fagnani F, et al. Aspartate aminotransferase and alanine aminotransferase activities in plasma: statistical distributions, individual variations, and reference values. Clin Chem 1975,21,1077–87.

[32] Schumann G, Bonora R, Ceriotti F, Ferard G, Ferrero CA, Franck PF, et al. IFCC primary reference procedures for the measurement of catalytic activity concentrations of enzymes at 37°C. International Federation of Clinical Chemistry and Laboratory Medicine. Part 5. Reference procedure for the measurement of catalytic concentration of aspartate aminotransferase. Clin Chem Lab Med 2002,40,725–33.

[33] Schumann G, Klauke R. New IFCC reference procedures for the determination of catalytic activity concentrations of five enzymes in serum: preliminary upper reference limits obtained in hospitalized subjects. Clin Chim Acta 2003,327,69–79.

[34] Ceriotti F, Henny J, Queralto J, Ziyu S, Ozarda Y, Chen B, et al. Common reference intervals for aspartate aminotransferase (AST), alanine aminotransferase (ALT) and γ-glutamyl transferase (GGT) in serum: results from an IFCC multicenter study. Clin Chem Lab Med 2010,48,1593–601.

[35] Kim HC, Nam CM, Jee SH, Han KH, Oh DK, Suh I. Normal serum aminotransferase concentration and risk of mortality from liver diseases: prospective cohort study. Br Med J 2004,328,983.

[36] Panteghini M, Bais R, van Solinge W. Enzymes. In: Burtis C, Ashwood E, Bruns D, eds. Tietz Textbook of Clinical Chemistry and Molecular Diagnostics, 4th edn. St Louis, MO, USA, Elsevier Saunders, 2006,597–644.

[37] Collier J, Bassendine M. How to respond to abnormal liver function tests. Clin Med 2002,2,406–9.

[38] Giannini EG, Testa R, Savarino V. Liver enzyme alteration: a guide for clinicians. CMAJ 2005,172,367–79.

[39] Goessling W, Friedman LS. Increased liver chemistry in an asymptomatic patient. Clin Gastroenterol Hepatol 2005,3,852–8.

[40] Hinson JA, Roberts DW, James LP. Mechanisms of acetaminophen-induced liver necrosis. Handb Exp Pharmacol 2010,196,369–405.

[41] Ellis G, Goldberg DM, Spooner RJ, Ward AM. Serum enzyme tests in diseases of the liver and biliary tree. Am J Clin Pathol 1978,70,248–58.

[42] Zimmerman HJ, Maddrey WC. Acetaminophen (paracetamol) hepatotoxicity with regular intake of alcohol: analysis of instances of therapeutic misadventure. Hepatology 1995,22,767–73.

[43] Dufour DR, Lott JA, Nolte FS, Gretch DR, Koff RS, Seeff LB. Diagnosis and monitoring of hepatic injury. II. Recommendations for use of laboratory tests in screening, diagnosis, and monitoring. Clin Chem 2000,46,2050–68.

[44] Pratt DS, Kaplan MM. Evaluation of abnormal liver-enzyme results in asymptomatic patients. N Engl J Med 2000,342,1266–71.

[45] Wai CT, Greenson JK, Fontana RJ, Kalbfleisch JD, Marrero JA, Conjeevaram HS, et al. A simple noninvasive index can predict both significant fibrosis and cirrhosis in patients with chronic hepatitis C. Hepatology 2003,38,518–26.

[46] Shaheen AA, Myers RP. Diagnostic accuracy of the aspartate aminotransferase-to-platelet ratio index for the prediction of hepatitis C-related fibrosis: a systematic review. Hepatology 2007,46,912–21.

[47] Manns MP, Czaja AJ, Gorham JD, Krawitt EL, Mieli-Vergani G, Vergani D, et al. Diagnosis and management of autoimmune hepatitis. Hepatology 2010,51,2193–213.

[48] Hennes EM, Zeniya M, Czaja AJ, Pares A, Dalekos GN, Krawitt EL, et al. Simplified criteria for the diagnosis of autoimmune hepatitis. Hepatology 2008,48,169–76.

[49] Alvarez F, Berg PA, Bianchi FB, Bianchi L, Burroughs AK, Cancado EL, et al. International Autoimmune Hepatitis Group Report: review of criteria for diagnosis of autoimmune hepatitis. J Hepatol 1999,31,929–38.

[50] Czaja AJ. Performance parameters of the diagnostic scoring systems for autoimmune hepatitis. Hepatology 2008,48,1540–8.

[51] Yeoman AD, Westbrook RH, Al-Chalabi T, Carey I, Heaton ND, Portmann BC, et al. Diagnostic value and utility of the simplified International Autoimmune Hepatitis Group (IAIHG) criteria in acute and chronic liver disease. Hepatology 2009,50,538–45.

[52] Das SK, Dhanya L, Vasudevan DM. Biomarkers of alcoholism: an updated review. Scand J Clin Lab Invest 2008,68,81–92.

[53] Cohen JA, Kaplan MM. The SGOT/SGPT ratio – an indicator of alcoholic liver disease. Dig Dis Sci 1979,24,835–8.

[54] Zhou SL, Gordon RE, Bradbury M, Stump D, Kiang CL, Berk PD. Ethanol up-regulates fatty acid uptake and plasma membrane expression and export of mitochondrial aspartate aminotransferase in HepG2 cells. Hepatology 1998,27,1064–74.

[55] Kew MC. Serum aminotransferase concentration as evidence of hepatocellular damage. Lancet 2000,355,591–2.

[56] McCullough AJ. The clinical features, diagnosis and natural history of nonalcoholic fatty liver disease. Clin Liver Dis 2004,8,521–33, viii.

[57] Torok NJ. Recent advances in the pathogenesis and diagnosis of liver fibrosis. J Gastroenterol 2008,43,315–21.

[58] Sorbi D, Boynton J, Lindor KD. The ratio of aspartate aminotransferase to alanine aminotransferase: potential value in differentiating nonalcoholic steatohepatitis from alcoholic liver disease. Am J Gastroenterol 1999,94,1018–22.

[59] Alexander J, Kowdley KV. HFE-associated hereditary hemochromatosis. Genet Med 2009,11,307–13.

[60] Castiella A, Zapata E, Otazua P, Fernandez J, Alustiza JM, Ugarte M, et al. [Utility of various non-invasive methods for fibrosis prediction among Basque Country patients with phenotypic hemochromatosis]. Rev Esp Enferm Dig 2008,100,611–4 (in Spanish).

[61] Beaton M, Guyader D, Deugnier Y, Moirand R, Chakrabarti S, Adams P. Noninvasive prediction of cirrhosis in C282Y-linked hemochromatosis. Hepatology 2002,36,673–8.

[62] Beaton M, Adams PC. Assessment of silent liver fibrosis in hemochromatosis C282Y homozygotes with normal transaminase levels. Clin Gastroenterol Hepatol 2008,6,713–4.

[63] Limdi JK, Hyde GM. Evaluation of abnormal liver function tests. Postgrad Med J 2003,79, 307–12.

[64] Giboney PT. Mildly elevated liver transaminase levels in the asymptomatic patient. Am Fam Physician 2005,71,1105–10.

[65] Eisenbach C, Sieg O, Stremmel W, Encke J, Merle U. Diagnostic criteria for acute liver failure due to Wilson disease. World J Gastroenterol 2007,13,1711–4.

[66] Korman JD, Volenberg I, Balko J, Webster J, Schiodt FV, Squires RH Jr, et al. Screening for Wilson disease in acute liver failure: a comparison of currently available diagnostic tests. Hepatology 2008,48,1167–74.

[67] Brancaccio P, Lippi G, Maffulli N. Biochemical markers of muscular damage. Clin Chem Lab Med 2010,48,757–67.

[68] Morse RP, Rosman NP. Diagnosis of occult muscular dystrophy: importance of the "chance" finding of elevated serum aminotransferase activities. J Pediatr 1993,122,254–6.

[69] Weibrecht K, Dayno M, Darling C, Bird SB. Liver aminotransferases are elevated with rhabdomyolysis in the absence of significant liver injury. J Med Toxicol 2010,6,294–300.

[70] Skenderi KP, Kavouras SA, Anastasiou CA, Yiannakouris N, Matalas AL. Exertional rhabdomyolysis during a 246-km continuous running race. Med Sci Sports Exerc 2006,38,1054–7.

[71] Carter JD, Kanik KS, Vasey FB, Valeriano-Marcet J. Dermatomyositis with normal creatine kinase and elevated aldolase levels. J Rheumatol 2001,28,2366–7.

[72] Mercado U. Dermatomyositis with normal creatine kinase and elevated aldolase levels. J Rheumatol 2002,29,2242, author reply 2242–3.

[73] Louthrenoo W, Weerayutwattana N, Lertprasertsuke N, Sukitawut W. Serum muscle enzymes, muscle pathology and clinical muscle weakness: correlation in Thai patients with polymyositis/dermatomyositis. J Med Assoc Thai 2002,85,26–32.

[74] Howie-Esquivel J, White M. Biomarkers in acute cardiovascular disease. J Cardiovasc Nurs 2008,23,124–31.

[75] Apple FS, Jesse RL, Newby LK, Wu AH, Christenson RH. National Academy of Clinical Biochemistry and IFCC Committee for Standardization of Markers of Cardiac Damage Laboratory Medicine Practice Guidelines. Analytical issues for biochemical markers of acute coronary syndromes. Circulation 2007,115,e352–5.

[76] Luepker RV, Apple FS, Christenson RH, Crow RS, Fortmann SP, Goff D, et al. Case definitions for acute coronary heart disease in epidemiology and clinical research studies: a statement from the AHA Council on Epidemiology and Prevention; AHA Statistics Committee; World Heart Federation Council on Epidemiology and Prevention; the European Society of Cardiology Working Group on Epidemiology and Prevention; Centers for Disease Control and Prevention; and the National Heart, Lung, and Blood Institute. Circulation 2003,108,2543–9.

[77] Panteghini M, Pagani F, Cuccia C. Activity of serum aspartate aminotransferase isoenzymes in patients with acute myocardial infarction. Clin Chem 1987,33,67–71.

[78] Rabkin SW, Desjardins P. Mitochondrial and cytoplasmic isoenzymes of aspartate aminotransferase in sera of patients after myocardial infarction. Clin Chim Acta 1984,138, 245–57.

[79] Annoni G, Chirillo R, Swannie D. Prognostic value of mitochondrial aspartate aminotransferase in acute myocardial infarction. Clin Biochem 1986,19,235–9.

[80] Freedman SB, Haladyn JK, Floh A, Kirsh JA, Taylor G, Thull-Freedman J. Pediatric myocarditis: emergency department clinical findings and diagnostic evaluation. Pediatrics 2007,120, 1278–85.

[81] Nanji AA. Markedly increased serum aspartate aminotransferase activity in congestive heart failure. Clin Biochem 1983,16,76–8.

[82] Tjon JM, van Bergen J, Koning F. Celiac disease: how complicated can it get? Immunogenetics 2010,62,641–51.

[83] Bardella MT, Vecchi M, Conte D, Del Ninno E, Fraquelli M, Pacchetti S, et al. Chronic unexplained hypertransaminasemia may be caused by occult celiac disease. Hepatology 1999,29,654–7.

[84] Tanner M, Kent N, Smith B, Fletcher S, Lewer M. Stability of common biochemical analytes in serum gel tubes subjected to various storage temperatures and times pre-centrifugation. Ann Clin Biochem 2008,45,375–9.

[85] Boyanton BL Jr, Blick KE. Stability studies of twenty-four analytes in human plasma and serum. Clin Chem 2002,48,2242–7.

[86] Bergmeyer HU, Bowers GN, Horder M, Moss DW. IFCC method for aspartate aminotransferase. Appendix B. Conditions for the measurement of the catalytic concentrations of reagent enzymes and the contaminants. Clin Chim Acta 1976,70,F41–2.

[87] Bergmeyer HU, Bowers GN Jr, Horder M, Moss DW. Provisional recommendations on IFCC methods for the measurement of catalytic concentrations of enzymes. Part 2. IFCC method for aspartate aminotransferase. Clin Chim Acta 1976,70,F19–29.

[88] Bergmeyer HU, Bowers GN, Horder M, Moss DW. IFCC method for aspartate aminotransferase. Appendix A. Description of pertinent factors in obtaining optimal conditions for measurement. Clin Chim Acta 1976,70,F31–40.

[89] Bergmeyer HU, Horder M, Rej R. International Federation of Clinical Chemistry (IFCC) Scientific Committee, Analytical Section. Approved recommendation (1985) on IFCC methods for the measurement of catalytic concentration of enzymes. Part 2. IFCC method for aspartate aminotransferase (L-aspartate: 2-oxoglutarate aminotransferase, EC 2.6.1.1). J Clin Chem Clin Biochem 1986,24,497–510.

[90] Panteghini M, Ceriotti F, Schumann G, Siekmann L. Establishing a reference system in clinical enzymology. Clin Chem Lab Med 2001,39,795–800.

[91] Schumann G, Bonora R, Ceriotti F, Clerc-Renaud P, Ferrero CA, Ferard G, et al. IFCC primary reference procedures for the measurement of catalytic activity concentrations of enzymes at 37°C. Part 2. Reference procedure for the measurement of catalytic concentration of creatine kinase. Clin Chem Lab Med 2002,40,635–42.

[92] Schumann G, Bonora R, Ceriotti F, Clerc-Renaud P, Ferrero CA, Ferard G, et al. IFCC primary reference procedures for the measurement of catalytic activity concentrations of enzymes at 37°C. Part 3. Reference procedure for the measurement of catalytic concentration of lactate dehydrogenase. Clin Chem Lab Med 2002,40,643–8.

[93] Schumann G, Bonora R, Ceriotti F, Ferard G, Ferrero CA, Franck PF, et al. IFCC primary reference procedures for the measurement of catalytic activity concentrations of enzymes at 37°C. International Federation of Clinical Chemistry and Laboratory Medicine. Part 6. Reference procedure for the measurement of catalytic concentration of γ-glutamyltransferase. Clin Chem Lab Med 2002,40,734–8.

[94] Schumann G, Bonora R, Ceriotti F, Ferard G, Ferrero CA, Franck PF, et al. IFCC primary reference procedures for the measurement of catalytic activity concentrations of enzymes at 37°C. International Federation of Clinical Chemistry and Laboratory Medicine. Part 4. Reference procedure for the measurement of catalytic concentration of alanine aminotransferase. Clin Chem Lab Med 2002,40,718–24.

[95] Siekmann L, Bonora R, Burtis CA, Ceriotti F, Clerc-Renaud P, Ferard G, et al. IFCC primary reference procedures for the measurement of catalytic activity concentrations of enzymes at 37°C. International Federation of Clinical Chemistry and Laboratory Medicine. Part 7. Certification of four reference materials for the determination of enzymatic activity of γ-glutamyltransferase, lactate dehydrogenase, alanine aminotransferase and creatine kinase accord. Clin Chem Lab Med 2002,40,739–45.

[96] Siekmann L, Bonora R, Burtis CA, Ceriotti F, Clerc-Renaud P, Ferard G, et al. IFCC primary reference procedures for the measurement of catalytic activity concentrations of enzymes at 37°C. Part 1. The concept of reference procedures for the measurement of catalytic activity concentrations of enzymes. Clin Chem Lab Med 2002,40,631–4.

[97] Schumann G, Canalias F, Joergensen PJ, Kang D, Lessinger JM, Klauke R, et al. IFCC reference procedures for measurement of the catalytic concentrations of enzymes: corrigendum, notes and useful advice. International Federation of Clinical Chemistry and Laboratory Medicine (IFCC) – IFCC Scientific Division. Clin Chem Lab Med 2010,48,615–21.

[98] Toussaint B, Emons H, Schimmel HG, Bossert-Reuther S, Canalias F, Ceriotti F, et al. Traceability of values for catalytic activity concentration of enzymes. a certified reference material for aspartate transaminase. Clin Chem Lab Med 2010,48,795–803.

[99] Lippi G, Salvagno GL, Montagnana M, Brocco G, Guidi GC. Influence of hemolysis on routine clinical chemistry testing. Clin Chem Lab Med 2006,44,311–6.

[100] Curry SC, Connor DA, Raschke RA. Effect of the cyanide antidote hydroxocobalamin on commonly ordered serum chemistry studies. Ann Emerg Med 1994,24,65–7.

[101] Beckerman N, Leikin SM, Aitchinson R, Yen M, Wills BK. Laboratory interferences with the newer cyanide antidote: hydroxocobalamin. Semin Diagn Pathol 2009,26,49–52.

6 Creatine kinase, isoenzymes, and isoforms

Alan H.B. Wu

6.1 Case studies

6.1.1 Patient A

A 35-year-old construction worker has a previous history of acute coronary syndromes. He is a smoker, slightly overweight, has a family history of premature heart disease (father died in his mid-40s), and has type 2 diabetes. Prior to his acute myocardial infarction (AMI), a routine physician check-up showed that his total cholesterol was 6.08 mmol/L (low risk <5.18 mmol/L), high-density lipoprotein (HDL) cholesterol 1.17 mmol/L (low risk >1.27 mmol/L), low-density lipoprotein (LDL) cholesterol 3.83 mmol/L (low risk <110 mg/dL), and high sensitivity C-reactive protein (hs-CRP) 7.3 mg/L (low risk <1 mg/L). He recovers from his initial event and is placed on β-blockers to improve his heart rhythm and a statin drug to reduce his lipids. Now after 6 weeks of lipid-lowering therapy, his lipid profile has improved with a total cholesterol 4.89 mmol/L, HDL 1.27mmol/L, LDL 2.41 mmol/L, and hs-CRP 1.5 mg/L. With regard to hepatic function, the patient shows no sign of liver toxicity with a mild increase in aspartate aminotransferase (AST) 63 U/L (AST), and normal for other tests, alanine aminotransferase (ALT) 35 U/L, total bilirubin 13.7 μmol/L, and alkaline phosphatase (ALP) 113 U/L. Unfortunately, he complains of aches and skeletal muscle pain. Although he normally has some soreness due to his occupation, the magnitude of these symptoms has led him to miss work. Total creatine kinase (CK) was measured and was 350 U/L. His doctor switched to a different lipid-lowering medication and decreased his dose. This resulted in the relief of his muscle pain and a return to his normal job duties.

6.1.1.1 Discussion

The statin drugs are 3-hydroxy-3-methyl-glutaryl-CoA (HMG-CoA) reductase inhibitors (rate-limiting step in the biosynthesis of cholesterol) and are widely used to treat patients with hyperlipidemia. Regular statin use reduces total and LDL cholesterol and inflammation as reflected by a decrease in hs-CRP. Although the majority of cardiac patients tolerate statins without any side effects, there are some patients who develop toxicities to the liver and skeletal muscles. Many physicians have advocated ordering liver function and CK tests at baseline, and within the first 3 and 12 months after the initiation of statins. Evidence of liver or skeletal muscle disease may result in a lowering of the statin dose or switching to an alternate medication. In this case, there was a mild elevation in CK consistent with the clinical history of myopathy. The increased

AST with normal results for the other liver function tests including ALT suggested that the AST increase was due to skeletal muscle damage. Skeletal muscle myopathy refers to the presence of any skeletal muscle diseases, which can have varying degrees of severity. The American Heart Association has identified three classifications of statin-induced muscle complaints. Myalgias include symptoms of pain without elevations in serum CK. Myositis is accompanied by increases of CK above three-fold the upper reference limit. The most severe case of statin-induced myopathy is rhabdomyolysis where the CK is greater than ten-fold the upper limit of normal [1]. The physician treating this case was appropriate in switching to a different statin and lowering the dose. It will be necessary to repeat blood tests for liver function and CK at 3 months to determine if this patient can tolerate this new prescription. The resolution of symptoms and return to employment suggested that this regimen was tolerated by the patient.

6.1.2 Patient B

A 40-year-old White alcoholic homeless male was found unconscious in an alley on a cold concrete pavement. According to his friends, he had been immobile for a period of 24 h. The patient was partially covered by newspapers, blankets, and cardboard. He has a history of cocaine and methamphetamine use. He was brought in by ambulance to the emergency department (ED). Because his body temperature was 35°C, a warming blanket was immediately applied to his body. Initial laboratory results showed a sodium level of 148 mmol/L, potassium 4.9 mmol/L, chloride 89 mmol/L, total CO_2 25 mmol/L, creatinine 88 μmol/L, blood urea nitrogen 5.4 mmol/L, CK 420 U/L, CK-MB 4.6 μg/L, relative index 1.1%, and myoglobin 32,000 μg/L. His urine was negative for myoglobin, positive for cocaine metabolites, and negative for the other drug classes tested (including amphetamines). Eight hours later, his temperature is normal at 37°C, renal function remains within normal limits, but his CK is now 11,350 U/L, CK-MB 136 μg/L, relative index 1.2%, and myoglobin 59,000 μg/L. On day 2, he is now mildly acidotic (pH 7.32), his CK has increased to 43,940 U/L, CK-MB 412 μg/L, relative index 0.9%, myoglobin 28,000 μg/L, creatinine has risen to 371 μmol/L, and potassium 5.3 mmol/L. He is diagnosed with acute renal failure induced by rhabdomyolysis and is treated with hemodialysis. After several days of hemodialysis, he recovers on day 6 from his renal failure, his CK activity and creatinine concentrations are normalized, and he is discharged to an outpatient facility for continuing care.

6.1.2.1 Discussion
This patient has developed rhabdomyolysis, an extensive and rapid breakdown of skeletal muscles, characterized by exceptionally high activities of CK, CK-MB, and myoglobin. The risk factors for rhabdomyolysis include immobilization, recreational drug use, statin drug use, trauma, extensive exertional exercise (e.g., marathon

running competitions), seizures, sepsis, viral illness, trauma, polymyositis, and genetic disorders. This patient exhibited several of the above named risk factors. His immobilization onto a cold surface caused muscle compression and poor circulation leading to injury. His electrolyte profile demonstrated dehydration, which is often seen in homeless individuals. A complication seen in patients with rhabdomyolysis is acute renal failure, as seen in this case. The mechanism of failure is related to overload of myoglobin, a low molecular weight protein (16.7 kDa) that is filtered by the glomerulus and causes obstruction within the renal tubules, and direct renal oxidative damage due to the presence of the heme group. In this case, the patient initially presented with normal renal function, as he was able to clear serum creatinine. However, he exhibited exceptionally high concentrations of CK and serum myoglobin that eventually led to overt failure. The absence of urine myoglobin suggested that renal failure was imminent due to the lack of renal clearance of this protein [2]. The hyperkalemia was due to the extensive skeletal muscle damage. Patients with rhabdomyolysis are treated with hydration and diuretics, and alkalinization. The use of sodium bicarbonate normalizes any metabolic acidosis present, while solubilizing myoglobin to hasten its clearance.

6.1.3 Patient C

A 96-year-old female is admitted to the ED with a 3-h history of chest pain. She has a previous history of coronary artery disease, hypertension, and hyperlipidemia. Her electrocardiogram was non-diagnostic for AMI. Her initial laboratory results were cardiac troponin I (cTnI) <0.04 µg/L, CK 93 U/L, CK-MB 2.3 µg/L, and relative index 2.4%. Eight hours after ED presentation, these values increased to 0.25 µg/L, 215 U/L, 8.3 µg/L, and 5.1%, respectively. Owing to her advanced age, she is not treated with a cardiac catheterization, and instead is admitted to the coronary care unit where she is given supportive care. The levels of her cardiac markers decline and CK returns to normal over the ensuing days. On hospital day 6, all of her biomarkers increase dramatically above baseline. She suffers a cardiac arrest on day 7 (Figure 6.1) and dies a few hours later.

6.1.3.1 Discussion
This patient suffered an AMI based on her clinical presentation of chest pain and the rise and fall of cTnI. A joint committee of cardiology professional groups has opined that in the context of myocardial ischemia, changes in troponin (T or I) concentration establishes the diagnosis of acute coronary syndromes [3]. The measurement of CK and CK-MB provide confirmatory information but is largely redundant in this case. As shown in Figure 6.1, although both cTnI and CK-MB are increased at 8 h, the magnitude of the increase (expressed as multiples of the upper reference limit) is higher for cTnI due to its higher tissue content. CK-MB returns to normal within

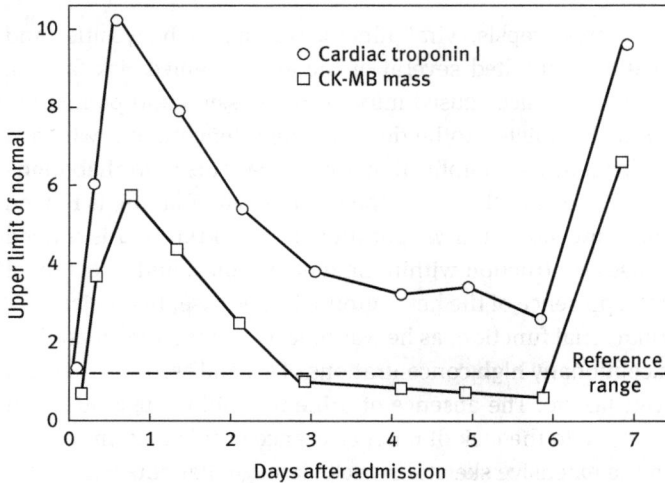

Figure 6.1: Cardiac biomarker results for Patient C. (o) Cardiac troponin I (cTnI). (□) CK-MB mass. Upper limit of normal is multiples of the upper reference range.

3 days but cTnI remains increased. This is due to the slow degradation of troponin from the myofibrils, unlike CK-MB which originates exclusively from the myocyte cytoplasm. Somewhere between hospital days 6 and 7, the patient suffers a reinfarction as signified by the increase for CK-MB and cTnI. This case illustrates the utility of cTnI for detection of reinfarction. Although cTnI does not return to baseline, the secondary rise is indicative of another ischemic episode. Apple and Murakami showed similar results to this patient in a series of confirmed myocardial reinfarction patients [4]. In most reinfarction cases, the second event is associated with a higher extent of myocardial damage and a secondary rise of cardiac biomarkers. Many clinical laboratories have begun to eliminate CK-MB testing for the diagnosis of AMI as a cost-savings measure. Opponents to this removal cite the need to retain CK-MB for detection of reinfarction. This and other cases show that troponin is equally effective in detecting this condition. It is important to note that although CK-MB may be eliminated for routine clinical laboratory practice, the measurement of total CK is needed for the evaluation of skeletal muscle disorders (see cases 1 and 2) and must be retained. Assays for total CK are considerably less expensive than for CK-MB and are readily available through automated clinical chemistry analyzers. If there is a desire or need to evaluate a patient for the presence of a reinfarction, the clinical sensitivity for use of total CK is equal to that of CK-MB.

6.1.4 Patient D

A 31-year-old elite female athlete experiences chest pain after running a marathon race. She had been intensively training for this and other long-distance races over the past several years. She is taken to a local hospital ED and arrives within 2 h after the completion of the race. The electrocardiogram and echocardiogram were both negative. At that time, her total CK was 7,200 U/L, CK-MB 364 µg/L, relative index 4.9%, cTnI <0.04 µg/L (<0.04 µg/L), and B-type natriuretic peptide (BNP) 35 ng/L (<100 ng/L). Based on these laboratory results, the patient is held overnight for observation. By the next morning her CK has risen to 10,500 U/L and CK-MB to 450 µg/L. However, her troponin is normal. She has no additional complaints and was discharged.

6.1.4.1 Discussion

Strenuous exercise can produce significant myocardial damage particularly in the untrained individual. In this case, the patient had experienced post-race chest pain that may have been consistent with myocardial injury. The total CK activity was incre-ased beyond the reference range, which was due to the exertional rhabdomyolysis, and is an expected finding for all participants of any marathon race. The CK-MB con-centration and relative index was also increased suggestive of a myocardial source. However, a normal cTnI and BNP suggested against a myocardial ischemic etiology as the source of CK-MB. Previous controlled studies have shown that there is a de-differentiation of skeletal muscle fibers from the predominately MM to the MB isoenzyme [5] resulting in a higher MB percentage in conditioned athletes relative to non-trained healthy controls. Metabolically active gastrocnemius muscle of long-distance runners appears to adapt to the heart muscle which may be more aero-bically efficient. As seen in this case, the higher CK-MB concentration was due to release from skeletal muscles that have been modified by training.

6.1.5 Patient E

A 44-year-old premenopausal female office executive experiences chest pain and is sent to the ED. She has no history of acute coronary artery disease. Although she has normal lipids, is normotensive, a non-smoker, no diabetes, and a nega-tive family history for premature cardiovascular disease, she is characterized as a "Type A" personality with much job-related stresses. In the ED, her CK was 96 U/L, CK-MB 3 µg/L, relative index 1%, hs-CRP 1.2 mg/L, and cardiac troponin 0.06 µg/L. A repeat cTnI at 4 h after presentation showed a result of 0.07 µg/L with continued normal results for CK and CK-MB. Based on these values, she was ruled out for an AMI and was discharged. She was referred to a cardiologist who subsequently scheduled her for an elective treadmill stress test the next morning. Shortly after the initiation of the test, she exhibited ST-segment depression on her Holter electrocardiogram and the test was prematurely stopped. She was immediately sent to the cardiac catheterization laboratory which revealed an 85% stenosis of her left anterior descending (LAD) coronary artery. The other arteries were less than 50% occluded. The LAD was opened with the catheter and a drug-eluting stent was inserted into this artery. She recovered without further incident and was discharged the following day.

6.1.5.1 Discussion

This case illustrates the value of cardiac troponin for risk stratification for future adverse cardiac events. The patient had normal values for CK and CK-MB and mild increases in cTnI. Repeat troponin revealed no significant increase from baseline value, effectively ruling out AMI in this case. However, a persistently mild increase in troponin in the context of myocardial ischemia put this patient at increased risk for AMI and/or cardiac death in the short term (e.g., 30 days). Troponin is a more sensitive marker than CK-MB for the detection of minor myocardial injury. Therefore, although both markers are generally equivalent for diagnosis and rule-out of AMI, troponin testing is superior for risk stratification. This is the reason why troponin has become the "gold standard" marker for acute coronary syndromes. This patient would have been sent home without any evidence of imminent acute cardiovascular disease if CK-MB testing was the only result available. She was negative for most of the traditional risk factors for atherosclerosis. The performance of the stress test, as suggested by her mild increase in cTnI, and the subsequent elective angioplasty may have averted her pending AMI.

6.2 Biochemistry and physiology

6.2.1 Molecular forms

CK (EC 2.7.3.2, adenosine triphosphate-creatine N-phosphotransferase), formerly known as creatine phosphokinase, is an important enzyme regulator of high-energy phosphate production and utilization. Although the highest concentration is seen in contractile tissue (skeletal muscle and myocardium), CK is also found in non-contractile organs such as the brain and distal renal tubules. CK has a dual function in the cells of these tissues. The mitochondrial CK form shuttles the production of ATP from aerobic metabolism to form creatine phosphate (CP), a reservoir for high-energy phosphate bonds (forward reaction 1). This intermediate is transported from the mitochondria to the cytoplasm where cytosolic CK converts CP back to ATP *in situ* for use by the cell (reverse reaction 1). For striated muscles, ATP is needed for contraction. In the brain and kidneys, CK is important to maintain electrolyte balance and provides the energy needed to maintain the sodium-potassium membrane pump. Equation (6.1) shows the reactions catalyzed by CK:

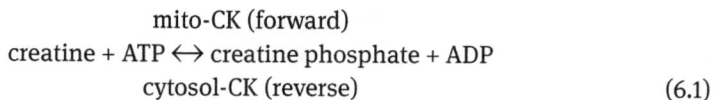

$$\text{mito-CK (forward)}$$
$$\text{creatine + ATP} \leftrightarrow \text{creatine phosphate + ADP}$$
$$\text{cytosol-CK (reverse)} \tag{6.1}$$

The forward reaction also recycles ADP within the electron transport system of the mitochondria, making it available as a phosphate acceptor molecule. This is more efficient than the slower diffusion of ADP back into the mitochondrial following consumption of ATP in the cytoplasm.

Cytosolic CK is encoded in chromosome 19 and is a dimer containing two distinct subunits, M and B. Each subunit has an active catalytic site, although the dimeric form is necessary for full catalytic activity. The combined molecular weight is approximately 86 kDa. Unlike myoglobin, this protein is too large to be filtered by the glomerulus; therefore, clearance from the circulation is dependent on the reticuloendothelial system.

6.2.2 CK isoenzymes, atypical forms, and isoforms

The combination of the two CK subunits produces the three major isoenzymes: CK-MM, CK-MB, and CK-BB. Figure 6.2 shows the tissue distribution of the cytosolic CK isoenzymes. As shown, the CK composition of skeletal muscles is largely the CK-MM isoenzyme, with 1–4% CK-MB. The total CK content is approximately 2,500 U/g wet weight. By contrast, the myocardium has a much higher CK-MB proportion, estimated to be between 15% and 40%. Nevertheless, the predominant form of CK in the heart is the MM isoenzyme, with a total CK content of 500 U/g wet weight. The CK-BB activity is the predominant CK isoenzyme found in the brain and smooth muscles with a content of 550 U/g wet weight. The CK-BB isoenzyme is also the major form found in the fetus and neonate. After birth, there is a steady conversion from CK-BB to the -MB and -MM isoenzymes within skeletal muscles. CK isoenzymes can be separated by agarose gel electrophoresis with the MM isoenzyme migrating to the cathode and the BB isoenzyme towards the anode (Figure 6.3).

In addition to the CK-MM, -MB, and -BB isoenzymes, there are also atypical isoenzymes that appear in blood on rare occasions. Macro-CK type 1 is an aggregate of CK-BB with immunoglobulin. The term is named because of the higher molecular weight that this complex exhibits. Binding to IgG is the most common among the macro-CK forms and migrates between the CK-MM and -MB isoenzymes by electrophoresis (Figure 6.3). When CK-BB binds to IgA, the migration is at or near that of CK-MB. Macromolecular forms have also been described for other enzymes such as amylase, lactate dehydrogenase, and AST. The mechanism for the formation of these variants is thought to be related to some autoimmune process. These enzymes are not cleared by the normal mechanism and are therefore characterized by persistently high concentrations.

Mitochondrial CK (mCK) also migrates before CK-MM and is found on rare occasions. There are two isoenzymes of mCK: sarcomeric, which occurs in striated muscle, and ubiquitous, which as the name implies, is found in nearly all tissues but especially the kidney and brain. In addition, mCK forms aggregates termed macro-CK type 2. This isoenzyme also migrates immediately cathodically to CK-MM. Dimeric mitochondrial CK is released in patients with malignant tumors. There is no clinical significance

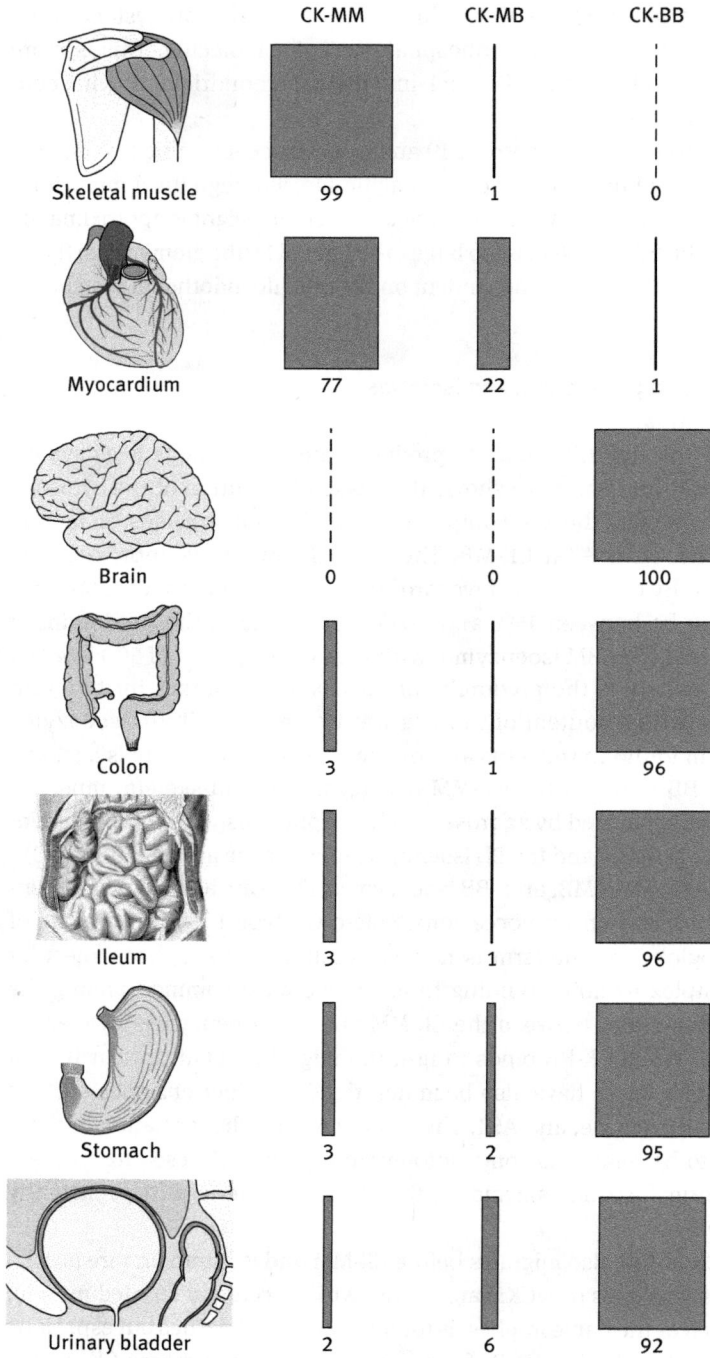

Figure 6.2: Tissue distribution of CK isoenzymes.

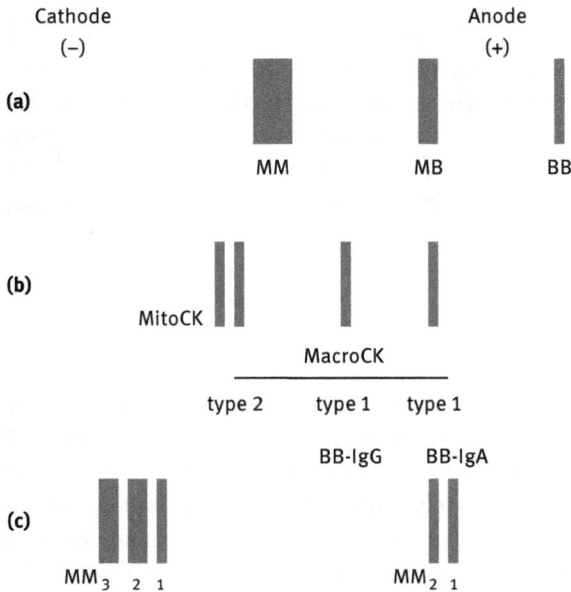

Figure 6.3: CK isoenzymes (a), variants (b), and isoforms (c) by electrophoresis.

towards the finding of macro-CK complexes. These variants are enzymatically active and add slightly to the total CK activity.

Similar to other macro-CK forms, the presence of mCK can cause an interference with CK isoenzyme activity assays, such as electrophoresis and immunoinhibition assays, but are not detected by CK-MB mass immunoassays. Using the latter technique, the presence of mCK and macro-CK forms can be readily discerned by two means. First, with serial testing, these abnormal forms result in persistently high apparent CK-MB activities, in contrast to a rising and falling pattern seen in acute coronary syndromes. Second, the relative index values of CK-MB/total CK activities are above and beyond the ratio seen in acute coronary syndromes and can even exceed 100% of the total CK activity. This is because the residual CK activity following immunoinhibition is multiplied by a factor of two, to account for each of the individually activity M and B subunits of the CK-MB isoenzyme. Without this correction factor, the true CK-MB activity would be 50% falsely low. In the presence of abnormal CK isoenzymes, none of the subunits are inhibited and the application of the correction factor simply doubles the degree of assay interference.

In contrast to the atypical isoenzymes, all patients undergo post-translational modifications of CK-MM and -MB to produce CK isoforms. Visualization of these forms requires use of high-resolution electrophoresis. The pure gene product is termed MM_3. Following release into serum, carboxypeptidase N convert the MM_3 isoenzyme to MM_2 through cleavage of lysine from the N-terminal end of one M subunit, and to MM_1 when the lysine is removed from the second M subunit. Two forms of CK-MB are found: MB_2, the gene product, and MB_1, the isoform where the M subunit has been modified. Figure 6.3c shows the relationship of the MM and MB isoforms. In normal blood, the concentration of the converted product MM_1 and MB_1 are the highest. However, when there is recent myocardial or skeletal muscle necrosis, there is a relatively higher concentration of the gene product, MM_3 and MB_2. At one time, the ratio of CK isoforms was used as an early indicator of myocardial infarction, but the test was never adopted.

6.2.3 Reference range

The reference range for total CK varies by gender, age, and ethnicity, and is largely based on muscle mass and degree of physical activity. Older subjects have reduced muscle mass and lower activities and therefore lower CK values. Men have higher reference ranges due to a higher skeletal muscle content. African Americans generally have higher values than Caucasians of the same age and gender. Clinical laboratories typically have gender-specific reference ranges but are not broken down further by age or ethnicity.

6.3 Chemical pathology

6.3.1 Heart disease

Prior to the development and implementation of assays for cardiac troponin, CK, and CK-MB were the standard biochemical marker for AMI. CK and CK-MB are also increased in patients after cardiac catheterization and cardiac surgery, valve disease, myocarditis, cardioversion, congestive heart failure, ventricular arrhythmias, and other cardiac-related problems. Chronic cardiac disease will not produce the same transient pattern of enzyme release or magnitude of enzyme elevations as what is seen in acute coronary syndromes. Therefore, serial testing is an important component for AMI diagnosis and to rule out AMI.

For AMI, irreversible myocyte injury produces an increase in both the total CK activity and the CK-MB concentration. Enzymes are released from the heart and appear into blood within the first 3–6 h after onset of chest pain, peak at 12–24 h, and return to normal within 2–3 days. In contrast to this, while cardiac troponin rises about the same time as CK-MB, it has a higher peak concentration due to higher tissue content, and remains increased for 5–7 days after the infarction. Although the early release of troponin is from the cytoplasm, the prolonged window of positivity for troponin is due to breakdown and release of troponin from the cytoplasmic and structural elements. CK is found in the cytoplasm alone and is therefore cleared more rapidly (Figure 6.4).

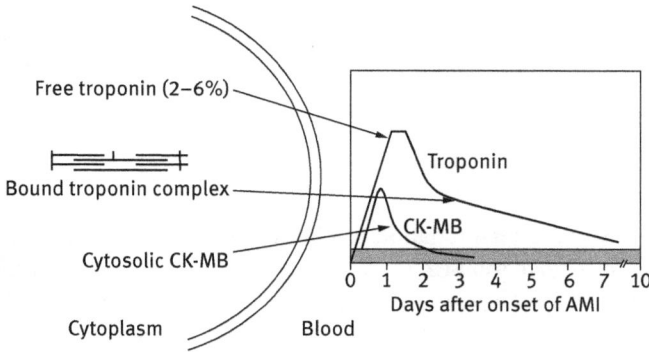

Figure 6.4: Release of CK-MB and troponin following acute myocardial infarction (AMI).

Despite its widespread use and success over many years, the main limitation of CK-MB as a biomarker for cardiac disease is the lack of specificity towards cardiac damage. Absolute increases in CK-MB can be observed in skeletal muscle injury due to the trace MB content found in muscle. Extensive skeletal muscle injury will greatly increase CK-MB concentrations in the absence of a myocardial source. Although the relative index will be normal in cases of skeletal muscle damage, some individuals have a higher CK-MB in their muscles (e.g., Patient D), and the relative index is increased even in the absence of heart disease. Testing for troponin is superior because of the absence of cardiac troponin within skeletal muscle tissue. Normal values of cardiac troponin are extremely low, enabling the use of low cut-off concentrations (i.e., the 99th percentile limit) for detection of minor myocardial injury. Use of this cut-off strategy can improve the risk stratification capability of troponin, as illustrated in Patient E. For CK-MB, the contribution of skeletal muscle isoenzyme adds significantly to baseline values. Cut-off concentrations are established empirically differentiating between minor myocardial injury as seen in unstable angina from myocardial infarction using receiver operating characteristic (ROC) curve analysis. Normal values for CK-MB were not used historically to establish AMI cut-off concentrations.

For these reasons, many laboratories have opted to eliminate CK-MB testing as a cost-savings measure, in favor of cardiac troponin in the diagnosis and management of AMI. In the majority of cases, CK-MB provides redundant and inferior clinical information and retention of this test has been to the appeasement of older physicians and cardiologists. There have been only a few arguments for the retention of CK-MB testing in conjunction with troponin. Some have argued that reinfarction can be more easily detected with CK-MB, as the concentration of the enzyme returns to baseline levels sooner. However, as shown in Patient C, troponin does show a secondary rise and is diagnostic of a reinfarction. Others have suggested that CK-MB rises ahead of troponin in an evolving AMI. With the use of high-sensitivity assays with the 99th percentile, studies have shown that troponin rises sooner than CK-MB and even myoglobin in some studies [6]. CK-MB is necessary for areas of the world where cardiac troponin testing is not available such as underdeveloped nations, owing to economic reasons.

6.3.2 Skeletal muscle disease

Although there is a trend towards the elimination of CK-MB, no such recommendations have or will be made to remove total CK from a laboratory's menu of offering. Measurement of this enzyme remains critically important in the assessment of skeletal muscle disease and trauma. Patients A and B illustrate the need for CK measurements in the evaluation of patients with myalgias and rhabdomyolysis. These conditions remain as common and significant causes of morbidity and mortality. The incidence of statin-induced myalgia ranges from 1% to 5% from clinical trials, but a higher incidence averaging 5–10% is seen in routine clinical practice [7]. Although the current target for LDL lowering is 100 mg/dL, the National Cholesterol Education Program has suggested further lowering to <70 mg/dL for extremely high-risk patients [8]. This will be achieved with increasing the statin dose, at the expense of a higher incidence of myalgias. The incidence of statin-induced rhabdomyolysis is rare, although one statin drug, cerivastatin (Baycol), was removed from the market.

CK is also increased in children with muscular dystrophies including Duchenne, Myotonic, Becker, and Limb-girdle. These genetic disorders are associated with very high CK activities. A child with muscular dystrophy usually has symptoms by the age of 5 years which include progressive muscle weakness and bilateral atrophy mostly to the legs and hips, but also the upper body. Most children are unable to walk before they reach teenage years. The diagnosis is based on physical examination, family history, electromyographic or nerve conduction tests, muscle biopsies, clinical laboratory tests, including CK, and genetic tests. Mutations in the dystrophin gene account for some of the muscular dystrophies.

6.3.3 Trauma

Patients undergoing any form of muscle trauma will have abnormal activities of CK, including motor vehicle accidents, extensive physical training involving muscle injury (running, marching, calisthenics, etc.), intramuscular injections, bee stings, spider bites, burns, electrical shock, and surgical procedures of any kind. Although absolute concentrations of CK-MB may be increased, the relative index (CK-MB/total CK) is within the rage expected for skeletal muscles, thereby establishing this as the source of enzyme release. For patients who may have also suffered acute coronary syndromes, the relative index for CK-MB will be negative and potentially misleading. Table 6.1 illustrates the inability of CK-MB to diagnose heart disease in the presence of concomitant skeletal muscle injury. As shown, the relative index of the

Table 6.1: Creatine kinase and troponin in acute coronary syndromes, muscle trauma, and combined disorder.

Disease or condition	Total CK U/L	CK-MB, ng/mL	Relative index, %	Troponin, ng/mL
Healthy	150	2	1.3	<0.04
Acute coronary syndromes	5,000	250	5.0	4.0
Trauma	10,000	100	1.0	<0.04
Combined	15,000	350	2.3	4.0

combined disorder is below the cut-off concentration because the CK-MB released from the myocardium will be obscured by the larger skeletal muscle component. Therefore, CK cannot be used to determine cardiovascular disease when there is also skeletal muscle release. If cardiac disease is to be assessed, cardiac troponin is the best serum marker as the cardiac isotype is not released following skeletal muscle injury. Table 6.1 shows that the troponin concentration is not influenced by skeletal muscle injury.

6.4 Analytical measurement of total CK and CK-MB

A standard assay for total CK has been established by the International Federation of Clinical Chemistry and has been adopted by essentially all manufacturers of CK reagents. It follows a coupled reaction where the rate of ATP production is rate limiting, Equation (6.2):

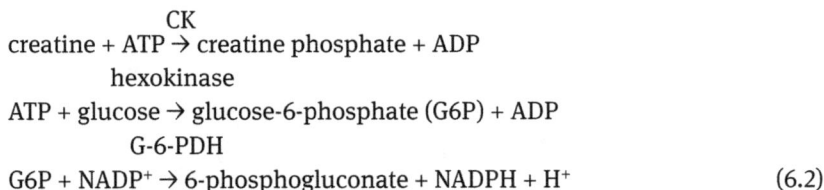

$$
\begin{aligned}
&\quad\quad\quad\text{CK} \\
&\text{creatine} + \text{ATP} \rightarrow \text{creatine phosphate} + \text{ADP} \\
&\quad\quad\text{hexokinase} \\
&\text{ATP} + \text{glucose} \rightarrow \text{glucose-6-phosphate (G6P)} + \text{ADP} \\
&\quad\quad\quad\text{G-6-PDH} \\
&\text{G6P} + \text{NADP}^+ \rightarrow \text{6-phosphogluconate} + \text{NADPH} + \text{H}^+ \quad\quad\quad (6.2)
\end{aligned}
$$

Although the standard assay temperature is 30°C, essentially all routine instruments measure this enzyme at 37°C, which results in higher enzyme activities. The reaction is monitored by an increase in absorbance at 340 nm due to the production of NADPH. Included in the reagent are magnesium as an activator and N-acetyl-L-cysteine or dithioerythritol as a stabilizer, and adenosine monophosphate as an inhibitor of adenylate kinase. There is a reasonably good standardization of CK results among manufacturers of reagents. In a recent proficiency testing survey for CK from the College of American Pathologists, the coefficient of variance for nearly 5,000 laboratories using 20 methods was 10%.

Historically, CK isoenzyme activities were measured using a variety of techniques including electrophoresis, column chromatography, and immunoinhibition assays. These have essentially been replaced by "sandwich" type immunoassays that are available on automated analyzers. Most assays use two antibodies, one that recognizes the individual B subunit and the other that recognizes the MB dimer [9]. One of these antibodies is conjugated to a bead or paramagnetic particle (capture), whereas the other is linked to some signal molecule (labeled), for example, a fluorescent, chemiluminescent, or electrochemical tag. The assay is performed by adding a serum sample to the mixture of antibodies. In theory, only the target analyte, MB isoenzyme, binds to both antibodies. The bound complex of antigen with antibodies is physically separated from the other proteins from the sample and unbound antibodies from the reagent. A measurement of the label is generated and compared against a calibration curve created from CK-MB standards. Although there is a reference material available from the National Institute on Standards and Technology, the standardization of CK-MB assays is incomplete with significant biases between manufacturers.

Sandwich immunoassays collectively suffer from the presence of heterophile and human anti-mouse antibodies (HAMAs) [10]. The presence of these atypical antibodies in patient serum samples causes false-positive results by binding to both the capture and labeled antibodies. There are several steps that laboratories can take to detect heterophile and HAMAs. Results are usually persistently high over time unlike samples from patients with acute diseases. There is a lack of recovery when these samples are diluted. Also, there is usually poor correlation of CK-MB results between different assays due to variances in the reagent formulation and specific antibodies used. If different CK-MB assays are available, these samples should be tested on these platforms. Commercial blocking agents are available and can be added to the suspected sample to eliminate their interference.

6.5 Questions and answers

1. The predominant form of CK found in the fetus is:
    (a) CK-MM
    (b) CK-MB
    (c) CK-BB
    (d) Macro-CK type 1
    (e) Macro-CK type 2
2. The clinical significance of macro-CK type 1 is:
    (a) Strong association with myocardial infarction
    (b) Strong association with risk of cardiovascular disease
    (c) Strong association with risk of stroke
    (d) No clinical significance
    (e) Strong association with diabetes

3. The CK isoform produced in tissue is termed:
   (a) CK-MM
   (b) $CK-MM_3$
   (c) $CK-MM_2$
   (d) $CK-MM_1$
   (e) $CK-MB_1$
4. Which of the following is the first to appear in serum following an AMI?
   (a) CK-BB
   (b) $CK-MM_3$
   (c) $CK-MM_2$
   (d) $CK-MM_1$
   (e) $CK-MB_1$
5. Which of the following is used to stabilize magnesium in assays used to measure CK activity?
   (a) Magnesium
   (b) Creatine
   (c) NADPH
   (d) N-acetyl-L-cysteine
   (e) Carboxypeptidase N

References

[1] Pasternak RC, Smith SC, Bairey-Merz CN, Grundy SM, Cleeman JI, Lenfant C. ACC/AHA/NHLBI clinical advisory on the use and safety of statins. J Am Coll Cardiol 2002,40,567–72.
[2] Wu AHB, Laios I, Green S, Gornet TG, Wong SS, Parmaley L, et al. Immunoassays for serum and urine myoglobin: myoglobin clearance assessed as a risk factor for acute renal failure. Clin Chem 1994,40,796–802.
[3] Thygesen K, Alpert JS, White HD, Jaffe AS, Apple FS, Galvani M, Joint ESC/ACCF/AHA/WHF Task Force for the Redefinition of Myocardial Infarction, et al. Universal definition of myocardial infarction. Circulation 2007,116,2634–53.
[4] Apple FS, Murakami MM. Cardiac troponin and creatine kinase MB monitoring during in-hospital myocardial reinfarction. Clin Chem 2005,51,460–3.
[5] Apple FS, Rogers MA, Sherman WM, Costill DL, Hagerman FC, Ivy JL. Profile of creatine kinase isoenzymes in skeletal muscles of marathon runners. Clin Chem 1984,30,413–6.
[6] Eggers KM, Oldgren J, Nordenskjold A, Lindahl B. Diagnostic value of serial measurement of cardiac markers in patients with chest pain: limited value of adding myoglobin to troponin I for exclusion of myocardial infarction. Am Heart J 2004,148,574–81.
[7] Thompson P, Clarkson P, Karas RH. Statin-associated myopathy. JAMA 2003,289,1681–90.
[8] Grundy SM, Cleeman JI, Bairey Merz CN, et al. Implications of recent clinical trials for the National Cholesterol Education Program Adult Treatment Panel III guidelines. Circulation 2004,110,227–39.
[9] Vaidya HC, Maynard Y, Dietzler DN, Ladenson JH. Direct measurement of creatine kinase-MB activity in serum after extraction with a monoclonal antibody specific to the MB isoenzyme. Clin Chem 1986,32,657–63.
[10] Kaplan IV, Levinson SS. When is a heterophile antibody not a heterophile antibody? When it is an antibody against a specific immunogen. Clin Chem 1999,45,616–8.

7 Gamma-glutamyl transferase

Sarah M. Brown

7.1 Case studies

7.1.1 Patient A

An 82-year-old retired female nurse with coronary artery disease (CAD) is seen by her cardiologist for routine visit. In addition to CAD, her past medical history is significant for multiple myeloma. She is on a variety of medications including lipid-lowering drugs, aspirin, and vitamin D. Her routine laboratory tests included a fasting lipid panel with a low-density lipoprotein (LDL) cholesterol of 1.89 mmol/L, high-density lipoprotein (HDL) cholesterol of 1.37 mmol/L, triglycerides of 1.16 mmol/L, and total cholesterol of 3.78 mmol/L. Her complete metabolic panel was unremarkable except for elevated alkaline phosphatase (ALP) of 188 U/L. A subsequent gamma glutamyl transferase (GGT) was mildly increased at 130 U/L. Abdominal ultrasound showed no significant abnormalities that could explain the test result. A bone survey showed some lucency in the femoral artery bilaterally, related to multiple myeloma.

7.1.1.1 Discussion

GGT synthesis is induced by hepatic microsomal enzymes and therefore serum levels will often be elevated due to prescription drugs. Some drugs notorious for causing elevated serum GGT are non-steroidal anti-inflammatory drugs (NSAIDs), antihypertensives, antidepressants, and anti-epileptics [1]. This patient is currently taking more than one of the drugs known to induce GGT synthesis, including an angiotensin-converting-enzyme (ACE) inhibitor and aspirin. The bone discrepancies related to her multiple myeloma explain her elevated ALP. An abdominal ultrasound to rule out hepatobiliary injury was unremarkable; the elevated GGT was attributed to her medications.

7.1.2 Patient B

A 46-year-old salesman is admitted to the alcohol detoxification ward following a 5-day drinking binge. He has a history of alcohol abuse, and this is his third admission for rehabilitation. He complains of abdominal pain, generalized weakness, and nausea. There is no history of intravenous drug abuse or blood transfusions. On physical examination, he is found to have considerable atrophy of the extremities, a protuberant abdomen, scleral icterus, and a low grade fever. Spider angiomata are noted on the skin of the thorax, as are the presence of hepatomegaly and tenderness of the liver to palpitation.

A complete blood count reveals a hemoglobin of 98 g/L, hematocrit of 30%, mean cell volume (MCV) of 100 fL, and mean cell hemoglobin concentration of 30 pg/cell. The leukocyte count and leukocyte differential count are normal. Admission laboratory data on serum are: bilirubin 83.8 μmol/L, total protein 67 g/L, albumin 23 g/L, globulins 44 g/L, and prothrombin time 18 s (control 12 s). Serum enzyme activities are: ALP 294 U/L, aspartate aminotransferase (AST) 396 U/L, alanine aminotransferase (ALT) 184 U/L, creatine kinase (CK) 145 U/L, and GGT 374 U/L. Viral hepatitis markers are all negative. A trace concentration of myoglobin is detected in the urine.

7.1.2.1 Discussion
This individual had markedly abnormal serum ALP, ALT, AST, and GGT values. GGT is present in high activity in the canalicular portion of the hepatocyte membranes, with lesser activities in the plasma membrane of the epithelial cells that line the bile ducts; it is the most sensitive of tests of alcohol abuse. Other indicators of liver dysfunction include the abnormal total and direct bilirubin, the decreased albumin, and the increased prothrombin time [2–4]. The patient had a reduced ability to synthesize albumin and coagulation proteins in adequate amounts. An MCV of above 100 fL is a common finding in chronic drinkers. A diagnosis of alcoholic hepatitis was made following a liver biopsy, which showed focal degeneration and necrosis of hepatic cells, acute inflammation, and alcoholic hyaline. Fatty change was also present. Viral hepatitis was ruled out based on the history and negative tests for viral hepatitis antigens and antibodies.

The abnormal CK, myoglobinuria, and skeletal muscle atrophy pointed to the presence of alcoholic myopathy in this patient. Alcohol has a toxic effect on skeletal muscle and myocardium, and the presence of even trace amounts of myoglobin in urine indicates active necrosis of skeletal and/or heart muscle. The man has a distinct risk of developing liver failure if he continues drinking.

7.1.3 Patient C

A fast-food restaurant employee in his twenties is admitted to the acute care unit following a motorcycle accident. He was thrown from the motorcycle and suffered head injury, blunt injury to the upper abdomen, and lacerations to his face, arms, and legs.

Physical examination shows a semicomatose man with bruises of his legs, arms, and forehead. Deep tendon reflexes are intact. The patient is unable to respond during the examination of the abdomen, and the extent of blunt injury cannot be determined. Radiographs show two fractured ribs and no apparent vertebral injuries. An abdominal computed chromatography (CT) shows no abnormalities.

Blood tests reveal a slightly decreased hemoglobin at 136 g/L and a leukocyte count of 15,000 cells/μL. He has normal arterial pH and blood gases. A blood alcohol test is negative and the serum osmolality is 300 mosmol. His CK is 910 U/K, CK-MB 2%, and lactate dehydrogenase (LD) 344 U/L. Other results are ALP 136 U/L, AST 74 U/L, and GGT 276 U/L.

The next day, the patient is slightly more alert. Hemoglobin and hematocrit values have not changed from the values obtained the previous day. A spinal tap is normal and no blood is present. On the third day, he is much more alert and complains of severe epigastric and back pain. The serum amylase is 580 U/L and a lipase 660 U/L.

7.1.3.1 Discussion

In addition to all skeletal muscle trauma he experienced, as reflected by the increased CK, LD, and AST, this man had blunt trauma to the abdomen and developed acute pancreatitis as indicated by his increased amylase and lipase activities in serum. The pancreatitis was of the mild, edematous type, and he never developed hemorrhagic necrosis of the gland. He had an uneventful recovery and was discharged after 1 week.

7.1.4 Patient D

A 59-year-old male chemist is admitted to the medical service with a history of recent progressive weight loss and poor appetite. The patient relates that he has lost approximately 14 kg in the past 6 months. His skin sags about his abdomen, and there is apparent atrophy of the muscles of the extremities. Ultrasound and CT scans reveal prominent enlargement of the body of the pancreas. A mass in the liver is detected at this time.

Laboratory data are: ALP 206 U/L, AST 86 U/L, LD 459 U/L, CK <25 U/L, carcino-embryonic antigen (CEA) 32 ng/mL, and GGT 644 U/L. A laparotomy is performed, and an inoperable carcinoma of the body of the pancreas with metastases to the liver is found.

7.1.4.1 Discussion

The diagnosis was made from the findings at laparotomy, although the abnormal CEA and greatly increased GGT suggested malignancy with possible liver involvement. Patients with cancer of the pancreas have a low 5-year survival, and often metastatic disease develops before symptoms appear. If the tumor compresses the common bile duct and extrahepatic cholestasis develops, it is sometimes possible to detect the malignancy before it is widespread, as seen in this patient.

7.1.5 Patient E

A 68-year-old fraternity housemother is admitted to the surgical service with symptoms of severe right upper quadrant abdominal pain with nausea and vomiting. Her symptoms first appeared 4 h after a heavy meal, and she has been ill for several days. She says her urine looks darker and her stool lighter than usual.

On physical examination, the patient is found to be obviously jaundiced. There is considerable tenderness in the right upper quadrant of the abdomen, and the liver is not palpable. Laboratory data obtained at this time are: serum total bilirubin 123 μmol/L, direct bilirubin 115 μmol/L, ALP 754 U/L, AST 116 U/L, CK 118 U/L, amylase 489 U/L, and GGT 654 U/L.

7.1.5.1 Discussion

This patient had a calculus in the biliary tract (choledocholithiasis), which became lodged in the biliary tract during the stimulation of bile flow by the fat-rich meal. Ultrasound examination revealed a gallstone near the ampulla of the common bile duct; the common hepatic and cystic ducts were dilated. The laboratory data reflected this dramatic extrahepatic jaundice with significantly increased serum direct and total bilirubin, ALP, and GGT. She had mild edematous pancreatitis, which resolved quickly. Apparently, this patient passed the stone on the third day of her hospitalization. A repeat ultrasound examination failed to detect a calculus. Her bilirubin returned to normal 2 weeks later.

7.2 Biochemistry and physiology

7.2.1 Molecular forms and post-translational modification

GGT, also known as gamma glutamyl transpeptidase, is a membrane-bound glycoprotein enzyme (EC 2.3.2.2) with multiple sites for carbohydrate binding. The amphophilic GGT polypeptide is cleaved into two fragments, a hydrophobic heavy chain that is anchored in the cell membrane and a hydrophilic light chain, which contains the active site and is extracellular [5]. In addition to the cell membrane, it is also present in the cytosol of some cells as a constituent of microsomes.

GGT does not have true isoforms; instead, it has forms that vary by carbohydrate content and structure. These modified forms can be separated by electrophoresis; however, all forms have the same catalytic activity and, despite numerous investigations, no diagnostic utility for differentiating the modified forms of GGT has been determined [5–7]. There is disagreement as to the number to GGT forms, and no uniform numbering system has been adopted. The migration pattern of GGT isoforms on gel electrophoresis varies from person to person, including which isoform is the most prominent.

Chromosomal analysis, genome sequencing, and microarray studies have identified a "GGT gene family" in the human genome, consisting of at least seven GGT genes in chromosome 22q11 and some related sequences that may be pseudogenes. However, only one full mRNA, corresponding to gene 6, is expressed. The expression of the GGT mRNA is regulated by multiple promoters, which is thought to play a role in tissue-specific expression [8–11]. Single-base mutations in the heavy chain have been identified, but no mutations have been found in the light chain. No mutations found to date have been shown to impact protein function [12].

7.2.2 Biochemical function

GGT catalyzes the transfer of a glutamyl group from a donor substrate to an acceptor, usually an amino acid or peptide. The most common natural substrate for GGT is the tripeptide glutathione (GSH) during the γ-glutamyl cycle (Meister cycle) [13], Equation (8.1):

$$\text{Glutathione + amino acid} \rightarrow \text{γ-glutamylamino acid + cysteinylglycine} \quad (8.1)$$

During the cycle, the γ-glutamyl residue of GSH is transferred to an accepter amino acid or peptide. So modified, the amino acid is transferred across the cell membrane via active transport and enters the cell amino acid pool for utilization. GSH, important for reductive chemistry and defense against oxidative stress, is regenerated during the cycle. GGT is also involved in detoxification processes through the excretion of GSH-xenobiotic conjugates. GGT cleaves the γ-glutamyl group from the GSH moiety, priming the conjugate for addition of an acetyl group, forming a mercapturic acid which can be excreted in the urine [14, 15].

7.2.3 Tissue activities and concentrations

GGT is mainly found in the membranes of cells that show high secretory or absorptive capacity such as the epithelial cells that line the biliary tract, hepatic canaliculi, proximal renal tubules, pancreatic acinar, pancreatic ducts, and intestinal brush border cells [7, 16–19]. The activities of GGT, in order of decreasing activity per gram of tissue, are in the kidney, liver, pancreas, and intestine [5, 19–21]. Measurable GGT activity has not been found in skeletal muscle or myocardium [22]. In normal serum, the GGT activity is much less than that in the kidney and practically all of the serum GGT derives from the liver. In serum, there is very little GGT from kidney, pancreas, or intestine [20, 23–25].

7.2.4 Metabolic clearance

Most of the serum GGT is removed from the blood by the liver and is excreted in bile; however, a small amount is catabolized by the kidneys. The serum half-life of GGT in humans is 7 to 10 days. In 32 alcoholic patients who had been abstinent for 8 weeks, the mean half-life was 28 days [26].

7.2.5 Reference intervals

Reference ranges for GGT must be viewed with caution, because many factors affect serum GGT activity. GGT in neonates up to 1 year of age is approximately five times the adult range; the values decline rapidly after birth and at approximately 6–8 months of age they are generally within the adult reference range [5, 27, 28]. Children older than 6 months have serum GGT activities that are rather constant, making GGT a better parameter than ALP for liver disease in children. ALP varies with bone development and could obscure liver disease [29]. After 60 years of age, GGT increases with age. Men have higher values than women, and obesity and even modest alcohol consumption (<30 mL whiskey/day) tend to increase serum GGT. Pregnant women often have slightly lower values than non-pregnant women of the same age [30, 31]. Serum GGT does not seem to be affected by physical activity, time of year, eating, or type of diet [32, 33]. Iatrogenic effects, such as endoscopic retrograde cholangiopancreatography, do not increase serum GGT [34]. Mild (<two-fold) elevations in GGT have been seen 24 h after laparoscopic cholecystectomy, presumably due to the impact of hemodynamic changes induced by intraperitoneal carbon dioxide insufflations on the liver.

Developing suitable reference ranges for the method and population being served is important if meaningful comparisons are to be made. Ideally, the population should contain a large group of individuals from all age groups and genders. "Social" drinkers should be excluded from the reference population or be asked to abstain for at least 2 weeks before blood is obtained [35, 36]. Considerable caution is necessary in selecting members of the normal cohort, because patient-provided information on personal drinking habits is notoriously unreliable.

7.3 Chemical pathology

7.3.1 Causes of increased or decreased concentrations or activities

GGT is sensitive to any type of hepatic insult, and its induction by a wide variety of drugs makes it difficult to interpret. GGT activity is elevated in nearly all instances of hepatobiliary dysfunction [5, 17]. It is also elevated in patients with infectious hepatitis and primary and secondary liver cancer. Although GGT is present at high concentration in the renal proximal tubule, no elevation of GGT activity in serum is seen in kidney disease. The mechanism behind the elevation of GGT is not fully understood. However, it is induced by certain drugs, such as steroids, and by ethanol [15, 30, 37]. GGT is elevated in fetal liver, presumably due to hepatocyte generation. Likewise, it has been suggested that the elevation seen in alcoholic cirrhosis could be due to the regeneration of hepatocytes. Recent data suggest that inflammation induces GGT expression via stimulation of the nuclear factor-κB signaling pathway by inflammatory cytokines with the subsequent recruitment of Sp1 and RNA polymerase II to the GGT promoter [12].

7.3.2 Diagnostic utility of GGT

7.3.2.1 Determining source of ALP
Unlike ALP, GGT is not elevated in osteoblastic disease. Thus, GGT can be used to distinguish whether bone or liver is the source of elevated ALP. Indeed, this is a common clinical use of GGT. In the pediatric population, GGT has an advantage over ALP for determining liver damage because there can be significant variation in ALP due to bone growth [29].

7.3.2.2 Hepatobiliary diseases
GGT is a sensitive, but not specific, marker of hepatobiliary disease and biliary obstruction; abnormal serum activities occur in 85% to 93% of cases [5]. In patients with developing cholestasis, the serum GGT increases before hyperbilirubinemia develops; a normal GGT is rare in a patient with cholestasis. The increase in bilirubin is largely of the conjugated form and a simultaneous increase in ALP is also common in these patients [38].

GGT is bound to the lipoprotein membrane of cells. For unknown reasons, the intracellular synthesis of GGT (and ALP) increases during cholestasis. Regurgitation of surface-active bile acids during cholestasis facilitates the solubilization of the enzyme from the cell membranes and release into the blood [39].

The magnitude of serum GGT activities is helpful in interpreting the values. Even a slight injury to the hepatobiliary system can increase serum GGT. Slight or moderately increased activities (two to five times the upper reference limit) are seen in liver diseases where hepatocellular damage is the main pathological lesion, for example, in viral hepatitis, chronic active hepatitis, and active cirrhosis. Activities above five times normal generally suggest the presence of cholestasis or a cholestatic component of the pathological process. Examples of the latter are drug-induced intrahepatic cholestasis, biliary cirrhosis, and extrahepatic biliary obstruction of any cause. Patients with protracted cholestasis occasionally show dramatic increases of GGT of 50 to 100 times normal.

In patients with an unexplained and markedly increased GGT, the possibility of primary or secondary hepatic malignancy must be considered. Patients with a neoplasm metastatic to the liver rarely have a normal serum GGT.

7.3.2.3 Alcoholism

Chronic alcoholism and various forms of alcoholic liver disease such as fatty liver, hepatitis, or cirrhosis can produce persistently increased serum activities. Abnormal GGT activities have been found in 75% of patients who consume alcohol every day [30]. An increased serum GGT is an early sign of moderate to excessive alcohol use; it does not necessarily mean that the liver is damaged or that the patient will have other abnormal liver function tests [40]. Chronic drinkers with otherwise normal liver function tests who then stopped drinking had GGT activities return to normal in 2 to 5 weeks [5]. A disadvantage of serum GGT for identifying alcohol abuse is that it may be too sensitive – even social drinkers tend to have abnormal values. Likewise, due to the low sensitivity for detection of alcohol abuse in females, GGT is not a good predictor of fetal alcohol syndrome.

The mechanism by which alcohol and other drugs increase serum GGT is believed to be induction of microsomal enzymes with enhanced synthesis of GGT, release of membrane-bound GGT aided possibly by bile acids in patients with cholestasis, and production of biliary stasis with regurgitation of GGT from the bile ducts [41, 42].

Evidence suggests that alcoholics are more sensitive to alcohol-induced hepatocyte changes, damage, or parenchymal lesions than non-drinkers. Abstaining alcoholics and non-drinkers were given a challenge dose of alcohol of 1 g/kg of body weight. An increase in serum AST, ALT, and GGT 24 h later was significantly greater in the alcoholics than in the non-drinkers [43, 44].

Once patients develop fatty liver, portal fibrosis, or cirrhosis, much higher and persistent serum GGT values are generally present. GGT cannot be used to determine the extent or stage of alcoholic liver disease [45]. Furthermore, there is no relationship between GGT activity and the quantity of alcohol that has been consumed or the duration of alcoholism [23].

7.3.2.4 Pediatrics

Owing to bone turnover during growth, there is significant variation in serum ALP in pediatric patients, making GGT a better marker for hepatobiliary disease in this population. Also, pediatric patients are less likely to be taking multiple drugs; if on medications, these are likely to be known. The degree of variation of serum GGT values from the reference limit can be a diagnostic indicator of some childhood biliary disorders, including extrahepatic biliary atresia, sclerosing cholangitis, and the three subtypes of progressive familial intrahepatic cholestasis (PFIC1–3). In extrahepatic biliary atresia, sclerosing cholangitis, and PFIC3, serum GGT is elevated 10 to 100 times over the upper limit of the reference range. In PFIC1 and PFIC2, serum GGT is normal or below the reference limit, discordant with the degree of cholestasis. GGT is also a useful marker for cholestatic liver dysfunction due to parenteral nutrition (PN) in children with intestinal disease. A recent study showed that 84% of PN-dependent children with short bowel syndrome had elevated GGT, whereas only 54% had elevated ALP [29].

7.3.2.5 Miscellaneous causes of abnormal serum GGT

Recent studies have indicated that high normal serum concentration of GGT is associated with type 2 diabetes mellitus, cardiovascular disease (CVD), cancer and cancer drug resistance, and exposure to environmental pollutants. However, the diagnostic utility of an elevated serum GGT in these instances or appropriate interventions have not yet been determined.

Patients with diabetes often have abnormal GGT activities. The cause may be induction of GGT synthesis by the liver. Recently, numerous prospective, cohort studies have shown that mildly elevated serum GGT, at the upper limit of reference of range, is an independent predictor of incipient type 2 diabetes in both men and women. Although the mechanism is not known, one hypothesis is that fat distribution around the liver induces GGT synthesis [46].

Following acute myocardial infarction, many patients have moderate increases in serum GGT. If the patient develops heart failure and circulatory disturbances leading to hypoxic liver damage, then GGT is released due to hypoxic injury to hepatocytes. In most patients where there is liver congestion and hypoxia following acute myocardial infarction, other enzyme changes such as increases in serum ALP, ALT, and LD-5 isoenzyme tend to confirm liver injury. Mildly elevated serum GGT has been associated with CVD independent of other risk factors such as alcohol intake, smoking, pre-existing ischemic heart disease, diabetes mellitus, lipid-lowering medications, blood pressure, total and HDL-cholesterol, glucose and pulmonary function. A 7-year change of >9.2 U/L was significantly associated with increased risk of CVD in men. This association is less pronounced in women [46]. Also, an association between C-reactive protein, a clinically recognized risk factor for CVD, and high normal serum GGT has been described [47, 48]. However, further research is needed to confirm elevated GGT as a CVD risk factor.

In acute pancreatitis, the pancreas is often edematous and enlarged, causing cholestasis and an elevation in serum GGT in some patients. Although GGT is present in the pancreas, there is no pathognomonic serum "pancreatic" GGT isoenzyme pattern in patients with pancreatic disease.

In addition to primary and secondary hepatic neoplasms, elevated serum GGT has been found in patients with cancers of other tissues, including colon, astrocytic glioma, soft tissue sarcoma, melanoma, and leukemia. The source of GGT in these cases is the tumor cells themselves; the mechanism of release of GGT from the cell membrane is not understood. Some suggest that GGT released from neoplasms exists in macromolecule complexes that can be distinguished by lipoprotein content [15]. However, more research in this area is needed.

The proximal renal tubules have very high GGT activities, but an increased serum GGT is not a common finding in renal diseases. Although slightly increased GGT activities have been observed in patients with nephrotic syndrome, renal failure, and post-transplantational rejection [48], there is very little information available regarding a role for GGT in prediction or prognosis of chronic kidney disease [46].

Abnormal GGT values have also been found in patients with hyperthyroidism, in some neurological conditions, in rheumatoid arthritis, and in patients with lung disorders [49, 50]. Why patients with these diseases have elevated GGT is still not completely understood. The current hypothesis is that the elevation is produced in response to a pro-oxidant state. The common theme among CVD, diabetes, arthritis, pulmonary dysfunction, cancer, and rheumatoid arthritis is an increase in reactive oxidant species. The oxidant burden is modulated by endogenous antioxidants including GSH, which is transferred by GGT [12, 46, 47, 51]. The oxidative stress link is supported by the elevation of GGT over AST, which is a marker of hepatic injury and has no role in redox chemistry [46].

7.3.2.6 Drugs that induce GGT synthesis

Because GGT is present in microsomes and is part of drug metabolism (detoxification), it is often elevated when a patient is on certain chemotherapies. These include sedatives such as phenobarbital, antidepressants, anticonvulsants such as phenytoin, antihyperlipidemic agents such as clofibrate, NSAIDs, ACE inhibitors, and estrogens that are present in some contraceptive pills. Also, recent studies indicate that serum GGT may be elevated following exposure to environmental pollutants [46].

7.3.2.7 Summary of diagnostic utility

A common clinical use of GGT is to determine whether an elevated ALP is due to a hepatobiliary or osteoblastic source. GGT is helpful in identifying patients with cholestasis of any cause. ALP and bilirubin must be determined at the same time before a diagnosis of cholestasis is made. In nearly all patients with cholestasis, serum GGT, ALP, and bilirubin are abnormally increased. Primary or secondary malignant liver neoplasms nearly always cause very high serum GGT activities. Persistently increased GGT values and/or increasing values suggest malignancy. GGT has distinct limitations in diagnosing the type of liver disease when abnormal values are found. Other procedures, complementary laboratory tests, biopsy, and so on are usually needed to make a definitive diagnosis.

GGT is a good substitute for ALP in the diagnosis of suspected liver disease when the cause of an increased ALP is unclear, that is, if there is possible concurrent bone disease or in pregnant women who usually have an increased ALP derived from placenta. GGT is normal in bone diseases and pregnancy, and an increased GGT nearly always means that some form of liver disease is present.

Single determinations of GGT are of limited value in following alcohol abusers, patients with cholestatic diseases, or other causes of abnormal values. This test has its greatest value when the changes are followed serially. Decreasing values after alcohol abuse suggest withdrawal of alcohol use or resolution of the cholestatic process. The magnitude of the increase of GGT is also helpful in some patients: primary or secondary malignant liver neoplasms tend to show very high values, whereas intrahepatic cholestasis and extrahepatic obstruction tend to show lesser increases.

7.3.2.8 GGT in urine

Urine contains thermally stable, low molecular weight, dialyzable that should be removed prior to analysis. GGT activity in normal urine is higher than serum; urinary GGT arises from the kidney and is not a part of the glomerular filtrate [21, 52]. Urine GGT is believed to originate from the normal turnover of renal tubular epithelial cells. Significant increases in urinary GGT have been observed in patients with pyelonephritis, renal tubular disorders, and in renal insufficiency [52]. Testing for GGT in urine may have some value in determining acute kidney transplant rejection. Urinary alcohol concentrations have been found to correlate with serum GGT activities in 42 patients with alcoholic liver disease [26].

7.4 Analysis

7.4.1 Specimen and stability

Serum is the specimen of choice. Heparin, citrate, fluoride, and oxalate interfere with the preferred method for GGT activity. Serum GGT is stable for at least 5 days at 4°C and for at least 3 months if frozen.

7.4.2 Preferred method for GGT activity

The procedure endorsed by the International Federation of Clinical Chemistry requires L-γ-glutamyl-3-carboxy-4-nitroanilide as the substrate for GGT (the γ-glutamyl donor), and glycylglycine as the γ-glutamyl acceptor. GGT catalyzes the transfer of the donor to the acceptor, and the assay readout is the measurement at a wavelength of 412 nm, which corresponds to the reaction product 5-amino-2-nitrobenzoate.

7.5 Questions and answers

1. Which of the following do not contain high activities of GGT?
    (a) Proximal renal tubule cells
    (b) Hepatic canaliculi
    (c) Intestinal brush border cells
    (d) Biliary epithelial cells
    (e) Renal parenchymal cells
2. Which of the following organs contain the highest GGT activity per gram of tissue?
    (a) Kidney
    (b) Liver
    (c) Pancreas
    (d) Intestine
    (e) Brain
3. A two-fold increase in serum GGT activity would most likely been seen in which of the following:
    (a) Viral hepatitis
    (b) Drug-induced intrahepatic cholestasis
    (c) Biliary cirrhosis
    (d) Extrahepatic biliary obstruction due to gallstones
    (e) Osetomalacia
4. Urine GGT activities are highest in which of the following conditions:
    (a) Pyelonephritis
    (b) Acute pancreatitis
    (c) Viral hepatitis
    (d) Intrahepatic biliary obstruction
    (e) Biliary cirrhosis

References

[1] Giannini E, Testa R, Savarino V. Liver enzyme alterations: a guide for clinicians. CMAJ 2005,172,367–79.
[2] Neuman T, Spies C. Use of biomarkers for alcohol use disorders in clinical practice. Addiction 2003,98,81–91.
[3] Sharpe P. Biochemical detection and monitoring of alcohol abuse and abstinence. Ann Clin Biochem 2001,38,652–64.
[4] Joelson B, Hultberg B, Alwmark A, Gullstrand P, Bengmark S. Total serum bile acids, gamma-glutamyl transferase, pre-albumin and tyrosine. Sensitive markers of hepatic dysfunction in alcoholic liver cirrhosis. Scand J Gastroenterol 1983,18,497–501.
[5] Whitfield JB. 2001. Gamma glutamyl transferase. Crit Rev Clin Lab Sci 2001,38,263–355.
[6] Kok PJMJ, Seidel B, Holtkamp Huisman J. A new procedure for the visualization of multiple forms of gamma-glutamyltransferase (GGT). Clin Chem Acta 1978,90,209–16.
[7] Tate SS, Meister A. Gamma-glutamyl transpeptidase: catalytic, structural, and functional aspects. Mol Cell Biochem 1981,39,357–68.

[8] Figlewicz DA, Delattre O, Guellaen G, Krizus A, Thomas G, Zucman J, et al. Mapping of human gamma-glutamyl transpeptidase genes on chromosome 22 and other human autosomes. Genetics 1993,17,299–305.

[9] Collins JE, Mungall AJ, Badcock KL, Fay JM, Dunham I. The organization of the gamma-glutamyl transferase genes and other low copy repeats in human chromosome 22q11. Genome Res 1997,7,522–31.

[10] Courtay C, Heisterkamp N, Siest G, Groffen J. Expression of multiple γ-glutamyltransferase genes in man. Biochem J 1994,97,503–8.

[11] Lieberman MW, Barrios R, Carter BZ, Habib GM, Lebovitz RM, Rajagopalan S. Gamma-glutamyl transpeptidase. What does the organization and expression of a multipromoter gene tell us about it functions? Am J Pathol 1995,147,1175–85.

[12] Mistry D, Stockley RA. Gamma-glutamyl transferase: the silent partner? COPD 2010,7, 285–90.

[13] Griffith OW, Bridges R, Meister A. Evidence that the γ-glutamyl cycle functions in vivo using intracellular glutathione: effects of amino acids and selective inhibition of enzymes. Proc Natl Acad Sci USA 1978,75,5405–8.

[14] Pompella A, Corti A, Paolicchi A, Giommarelli C, Zunino F. γ-Glutamyltransferase, redox regulation and cancer drug resistance. Curr Opin Pharmacol 2007,7,360–6.

[15] Corti A, Franzini M, Paolicchi A, Pompella A. γ-Glutamyltransferase of cancer cells at the crossroads of tumor progression, drug resistance and drug targeting. Anticancer Res 2010,30,1169–82.

[16] Horiuchi S, Inoue M, Morino Y. γ-Glutamyl-transpeptidase. Sidedness of its active site of renal brush-border membrane. Eur J Biochem 1978,87,429–37.

[17] Javitt NB. Hepatobiliary disease. Annu Rev Clin Biochem 1980,1,93–138.

[18] Meister A, Tate SS, Ross LL. Membrane-bound gamma-glutamyltransferase. In: Martonosi A, ed. The enzymes of biological membranes, New York, NY, USA, Plenum Press, 1976,3,315–47.

[19] Miura T, Matsuda Y, Tsui A, Katunuma N. Immunological cross-reactivity of gamma-glutamyl transpeptidase from human and rat kidney, liver, and bile. J Biochem (Tokyo) 1981,89,217–22.

[20] Huesby NE. Multiple forms of gamma-glutamyl transferase in normal human liver, bile, and serum. Biochim Biophys Acta 1978,483,46–56.

[21] Linder M, Sudaka P. Etudes des formes d'elimination urinaire de la gamma-glutamyl transpeptidase et de l'aminopeptidase. Clin Chim Acta 1982,118,77–85.

[22] Lesgourgues B, Nalpas B, Berthelot P. Gamma-glutamyl-transferase: un test simple, une interpretation delicate. Gastroenterol Clin Biol 1984,8,99–102.

[23] Goldberg DM. Structural, functional, and clinical aspects of gamma-glutamyltransferase. CRC Crit Rev Clin Lab Sci 1980,12,1–58.

[24] Shaw LM, Peterson-Archer L, London JW, Marsh E. Electrophoretic, kinetic and immunoin-hibition properties of gammaglutamyltransferase from various tissues compared. Clin Chem 1980,26,1523–7.

[25] Rosalki SB. Gamma-glutatmyl transferase. Adv Clin Chem 1975,17,53–107.

[26] Orrego H, Blake JE, Israel Y. Relationship between gamma-glutamyl transpeptidase and mean urinary alcohol levels in alcoholics while drinking and after alcohol withdrawal. Alcohol Clin Exp Res 1985,9,10–3.

[27] Knight JA, Haymond RE. Gamma-glutamyl transferase and alkaline phosphatase activities compared in serum of normal children and children with liver disease. Clin Chem 1981,27, 48–51.

[28] Priolisi A, Ditana M, Fazio M, Gioeli RA. Variation of serum gamma-glutamyltransferase in full-term and pre-term babies during their first two weeks of life. Minerva Pediatr 1980,32,291–6.

[29] Cabrera-Abreu JC, Green A. Gamma-glutamyltransferase: value of its measurement in paediatrics. Ann Clin Biochem 2002,39,22–5.

[30] Rosalki SB, Rau D. Serum gamma-glutamyl transpeptidase activity in alcoholism. Clin Chem Acta 1972,39,41–7.
[31] Schiele F, Guilmin A-M, Detienne H, Siest G. Gamma-glutamyltransferase activity in plasma: statistical distributions, individual variations and reference intervals. Clin Chem 1977,23,1020–8.
[32] Haralambie G. Serum gamma-glutamyl transpeptidase and physical exercise. Clin Chim Acta 1976,72,363–9.
[33] Jacobs WL. Gamma-glutamyl-transpeptidase in diseases of the liver, cardiovascular system and diabetes mellitus. Clin Chim Acta 1972,38,419–34.
[34] Nemesanszky E, Tulassay Z, Papp J. Serum enzyme changes after endoscopic retrograde cholangio-pancreatography (ERCP). Acta Hepato-Gastroenterol 1978,25,228–32.
[35] Henny J, Siest G, Schiele F, Stein JM. Use of a reference state concept for interpretation of laboratory tests: drug effects on gamma-glutamyl transferase. Adv Biochem Pharmacol 1982,3,209–13.
[36] Siest G, Herbeth B, Schiele F, Henny J. References et variations biologigue de la gamma-glutamyltransferase. Bull Soc Dr Alcool 1983,5,13–20.
[37] Nemensanszky E. Induction of gamma-glutamyltransferase in alcoholics versus normal individuals. Clin Chem 1988,34,525–7.
[38] Rolsaki SB, Foo AY, Nemesansky E. Gamma-glutamyltranspeptidase isoenzymes in hepato-biliary diseases. Adv Biochem Pharmacol 1982,3,153–8.
[39] Ratanasavanh D, Tazi A, Gaspart E. Hepatic gamma-glutamyl transferase release. Effect of bile salt and membrane structure modifications. Advan Biochem Pharmacol 1982,3,93–103.
[40] Weil J, Schellenberg F, Legoff AM, Lamy J. The predictive value of gamma-glutamyltransferase and other peripheral markers in the screening of alcohol abuse. Adv Biochem Pharmacol 1982,3,195–8.
[41] Barouki R, Chobert MN, Finidori J, Aggerbeck M, Nalpas B, Hanoune J. Ethanol effects in rat hepatoma cell line: induction of gamma-glutamyltransferase. Hepatology 1983,3,323–9.
[42] Halsall S, Peters TJ. Effect of chronic ethanol consumption on the cellular and subcellular distribution of gamma-glutamyltransferase in rat liver. Enzyme 1984,31,221–8.
[43] Nemesanszky E, Lott JA. Gamma-glutamyltransferase and its isoenzymes: progress and problems. Clin Chem 1985,31,797–803.
[44] Shaw LM. Keeping pace with a popular enzyme: GGT. Diagn Med 1982,5,59–78.
[45] Teschke R, Rauen J, Stromeyer G. Alcoholic liver disease. Assessment of various stages by determination of the adult and fetal form of gamma-glutamyltransferase activity in serum. INSERM 1980,96,215–20.
[46] Targher G. Elevated serum γ-glutamyltransferase activity is associated with increased risk of mortality, incident type 2 diabetes, cardiovascular events, chronic kidney disease, and cancer – a narrative review. Clin Chem Lab Med 2010,48,147–57.
[47] Turgut O, Yilmaz A, Yalta K, Karadas F. γ-Glutamyltransferase is a promising biomarker for cardiovascular risk. Med Hypotheses 2006,67,1060–4.
[48] Penn R, Worthington DJ. Is serum gamma-glutamyltransferase a misleading test? Br Med J 1983,286,531–5.
[49] Spooner RJ, Smith DH, Bedford D, Beck PR. Serum gamma-glutamyltransferase and alkaline phosphatase in rheumatoid arthritis. J Clin Pathol 1982,35,638–41.
[50] Barton AP, Powers JL, Lourenco RV. Gamma-glutamyltranspeptidase in chronic obstructive pulmonary disease. Proc Soc Exp Biol Med 1974,146,99–103.
[51] Horpacsy G, Zinsmeyer J, Schroeder K, Mebel M. Value of determining urinary enzymes after human kidney transplantation. Early warnings of rejection or not? Clin Chem 1977,23,770–1.
[52] Salgo L, Szabo A. Gamma-glutamyl transpeptidase activity in human urine. Clin Chim Acta 1982,126,9–16.

8 Lactate dehydrogenase

Olajumoke Oladipo and Dennis J. Dietzen

8.1 Case studies

8.1.1 Patient A

A 66-year-old Caucasian female presented to the emergency room with a history of vomiting, diarrhea, worsening shortness of breath, and decreased oral intake. She was previously diagnosed with stage IV non-small cell lung carcinoma. She has a 20 pack-year history of cigarette smoking (1/2 a pack per day for 40 years) prior to diagnosis. On examination she had positive rhonchi bilaterally and decreased breath sounds in the right lower lobe. She had a right pleural effusion on chest X-ray. Her initial biochemical and hematology results are: serum chemistry – sodium 138 mmol/L, potassium 6.9 mmol/L, chloride 89 mmol/L, CO_2 15 mmol/L, glucose 8.1 mmol/L, blood urea nitrogen (BUN) 35.3 mmol/L, creatinine 301 µmol/L, phosphate 1.91 mmol/L, total calcium 2.2 mmol/L, albumin 26 g/L, alanine aminotransferase (ALT) 117 IU/L, alkaline phosphatase (ALP) 431 IU/L, lactate dehydrogenase (LD) 4,477 IU/L, total bilirubin 65 µmol/L, white blood cell (WBC) count 21.6×10^9/L, hemoglobin 143 g/L, mean cell volume (MCV) 98.4 fL, platelets 169×10^9/L, lymphocytes 5%, granulocytes 89.7%, monocytes 3.0%, basophils 1.9%, eosinophils 0.4%.

8.1.1.1 Discussion
This patient was found to have advanced carcinoma of the lungs and was admitted with a diagnosis of acute renal failure and metabolic acidosis secondary to vomiting and diarrhea. She also had pleural effusion for which she had a chest tube placed. Her liver enzymes showed that LD activity was increased to a greater degree compared with ALT and aspartate aminotransferase (AST) activities. The tumor cells were most likely the main source of LD. Some rapidly growing cancer cells depend on anaerobic glycolysis to generate ATP, even in the presence of adequate oxygen (the Warburg effect), thereby producing more lactate via LD [1]. Adenocarcinoma of the lung is strongly associated with cigarette smoking. A very high total LD, which can be as high as 20 times the upper reference limit, has been reported to be associated with poor outcome and a low survival rate [2]. Increased LD in lung carcinoma has been mainly attributed to LD_3, LD_4, and LD_5 isoenzymes [3, 4]. In this case, LD isoenzymes were not measured. The renal function of this patient was compromised, as indicated by abnormal BUN and creatinine concentrations. In addition, leukocytosis suggests that this patient has an underlying infection. All these processes would result in tissue destruction and contribute to the circulating pool of LD. Patients who have

higher serum LD are also at increased risk for developing post-operative morbidity [5]. Spontaneous tumor lysis syndrome has been reported in solid tumors, even though it is more common in patients on chemotherapy [6]. Her medical records do not list any form of chemotherapy treatment nor do her laboratory findings suggest tumor lysis syndrome, as she exhibits mild hyperphosphatemia probably due to metabolic acidosis and renal failure, normal calcium, and magnesium concentrations. Total LD activity may be a useful tool for follow-up in patients to monitor response to therapy. This patient was rehydrated and acute renal failure resolved; she was then discharged to hospice care.

8.1.2 Patient B

A 17-year-old G1 P0 female patient presented with left flank pain of approximately 1 month duration at 28 weeks' gestation. There was no history of hematuria, dysuria, or oliguria. She was afebrile with normal blood pressure. She had a gravid uterus and tenderness of the left flank. She was diagnosed with renal abnormality a year previous; however, she did not seek further treatment at this time. Abdominal ultrasound showed a solid heterogeneous large mass in the upper pole of the left kidney measuring 10.9 × 8.9 × 11.4 cm with blood flow. The right kidney showed some hydronephrosis. Magnetic resonance imaging identified lesions in the liver suggesting metastasis from the renal mass. A computed tomography scan confirmed that the lesion did not extend into the renal veins. Also, there was a nodule in the right lower lung. The spleen, gall bladder, and pancreas were unremarkable. Her serum laboratory results were: sodium 138 mmol/L, potassium 4.6 mmol/L, total protein 85 g/L, albumin 3.4 g/L, uric acid 434 µmol/L, total bilirubin 7.0 µmol/L, LD 1,500 IU/L, ALT 187 IU/L, AST 179 IU/L.

8.1.2.1 Discussion
This patient subsequently had a left nephrectomy a week after delivery of a healthy baby. Histological diagnosis of the mass showed it to be Wilms' tumor. The renal pathology was present before pregnancy but was not followed up and the physiological changes associated with pregnancy contributed to aggravating the disease process. The patient was induced at 34 weeks' gestation and had a left nephrectomy after an uneventful postnatal period. The liver biopsy confirmed metastasis. Wilms' tumor, an embryonic neoplasm, is the most common childhood renal tumor, accounting for approximately 6% of all pediatric malignancies. On rare occasions, Wilms' tumor has been described in teenagers and adults with worsening prognosis when compared with children [7, 8]. This patient had an LD of 1,500 IU/L with evidence of metastasis to the liver and possibly the lungs. At discharge following debulking the tumor, her LD decreased to 450 IU/L. Serum LD and its isoenzymes, LD_4 and LD_5, have

been proposed as tumor markers for either differential diagnosis or monitoring the response of treatment in patients with Wilms' tumor [9].

8.1.3 Patient C

A 34-year-old female with 33 weeks' twin pregnancy was admitted with a diagnosis of pre-eclampsia and growth discordance between the twins. On admission, she was hypertensive and had developed hyperuricemia. She also had proteinuria and hematuria. She was induced within 48 h of admission and delivered healthy twin boys. In the postpartum period, she had a perineal hematoma evacuated and subsequently developed a fever and abdominal tenderness. She was oliguric and had an enlarged liver on examination. Her laboratory investigations revealed increasing liver enzymes, decreasing platelets, and mild anemia.

Laboratory values measured in serum 24 h postpartum were: sodium 130 mmol/L, potassium 4.3 mmol/L, CO_2 22 mmol/L, chloride 103 mmol/L, uric acid 523 µmol/L, ALT 3,029 IU/L, AST 2,751 IU/L, LD 3,098 IU/L, total calcium 1.9 mmol/L, hemoglobin 113 g/L, WBC 33.2×10^9/L with 85% granulocytes, platelets 110×10^9/L, D-dimer 13,823 µg/L, fibrinogen 9.9 µmol/L.

8.1.3.1 Discussion
A diagnosis of HELLP syndrome (*H*emolysis, *E*levated *L*iver enzymes, and *L*ow *P*latelet count) was made and she was managed accordingly. By the 4th day postpartum, her liver enzymes and platelet count began to return to normal. She was discharged home 2 weeks after admission and LD and liver enzymes were within the reference intervals.

HELLP syndrome is a pregnancy associated thrombotic microangiopathic state characterized by hemolysis, elevated liver enzymes, and low platelet counts. An increased risk of developing HELLP has also been associated with LCHAD (*L*ong *C*hain 3-*H*ydroxyacyl-Co*A* *D*ehydrogenase) deficiency in the fetus [10]. This patient had pre-eclampsia which quickly progressed to HELLP, as evidenced by the rapidly increasing liver enzymes, decreasing platelet counts, and increased LD due to hemolysis. The activity of LD in the red blood cells is greater than 100 times that in serum; therefore, it is a very sensitive indicator of hemolysis. Her total bilirubin level was high, with a majority being unconjugated, which is another indicator of hemolysis. An increase in both AST and ALT is indicative of liver damage, even though ALT is regarded as the more specific liver enzyme. The very high D-dimer concentration is indicative of the thrombotic state. Serial measurements of the enzymes and bilirubin proved to be a very useful tool in predicting recovery in this patient. LD has been reported to be a very useful parameter used to predict recovery and progression of disease in these patients [11, 12]. As the patient's condition improved, the platelet count increased and the liver enzymes returned to within the reference intervals. The activity of LD decreased with recovery to 156 IU/L at discharge.

8.1.4 Patient D

A 70-year-old male patient presented to the clinic with malaise and loss of appetite. He also complained of increasing night sweats that required him to change his bed linens frequently. On examination, the physician noted non-tender lymph nodes in his neck, which were later confirmed by biopsy to harbor non-Hodgkin's lymphoma (NHL). His past medical history included non-insulin-dependent diabetes mellitus, gastroesophageal reflux, and chronic obstructive pulmonary disease.

Pre-chemotherapy serum biochemical investigations were: sodium 139 mmol/L, potassium 3.8 mmol/L, chloride 96 mmol/L, CO_2 32 mmol/L, total calcium 2.5 mmol/L, total protein 44 g/L, glucose 11.2 mmol/L, BUN 6.4 mmol/L, creatinine 75 μmol/L, ALP 215 IU/L, LD 850 IU/L, ALT 12 IU/L, AST 31 IU/L.

8.1.4.1 Discussion

Follicular lymphoma constitutes approximately 20% of all newly diagnosed lymphoma cases and is the second most common subtype worldwide. This patient had an elevated LD at diagnosis and studies have shown that increased LD at diagnosis correlates with lower response to therapy and shorter survival [13, 14]. The other co-morbidities demonstrated in this patient, hyperglycemia and hypoproteinemia, are also poor prognostic indicators. LD activity may be determined as a measure of tumor cell proliferation and for prognostic purposes in patients with NHL [15]. The most commonly used index for determining prognosis in patients with follicular lymphoma is the Follicular Lymphoma International Prognostic Index. The original version included LD activity as one of the indices, the others being age, tumor burden (Ann Arbor stage, number of nodal sites), and hemoglobin [16]. A revision of this index did not include LD in the criteria and this may be related to the fact that even though it is a robust prognostic factor in several lymphomas, it was found to be elevated in only 20% of patients with follicular cancer [17]. In patients with an elevated LD, it is expected that as the tumor burden reduces the LD will also fall, although there may be an initial increase due to tumor lysis by chemotherapy. The analysis of LD isoenzyme profiles in patients with hematopoietic malignancies typically show increased percentages of LD_2 in patients with NHL, chronic lymphocytic leukemia, and myeloproliferative syndromes, but not in samples from patients with multiple myeloma or Hodgkin's disease. High LD_3 values were predictive of early death in NHL patients (<2 months compared with 12 months in patients with normal LD_3) [18]. LD isoenzymes are not routinely measured in clinical laboratories and this patient did not have isoenzymes analyzed. The normal ALT and AST levels in this patient indicate that LD is not derived from the liver. The patient was started on chemotherapy and LD and liver enzymes were used for monitoring.

8.1.5 Patient E

An 11-year-old Caucasian female 3 years post-bone marrow transplant from an unrelated donor for acute leukemia presented with left upper quadrant pain and splenomegaly. She was maintained on a therapeutic regimen that included prednisone, tacrolimus, penicillin, folate, magnesium oxide, and acyclovir. Additional history includes chronic graft versus host disease (GVHD) and recent laparoscopic cholecystectomy. Results of laboratory tests performed at initial presentation were: sodium 143 mmol/L, potassium 4.1 mmol/L, chloride 106 mmol/L, total CO_2 26 mmol/L, AST 47 IU/L, ALT 32 IU/L, total bilirubin 149 μmol/L, haptoglobin <5 g/L, LD 1,465 IU/L, WBC 5.3×10^9/L, red blood cell count 2.7×10^{12}/L, platelets 91×10^9/L, hemoglobin 79 g/L, hematocrit 23%, direct Coombs – positive.

8.1.5.1 Discussion

This patient underwent a laparoscopic splenectomy for sequestration secondary to chronic autoimmune hemolytic anemia (AIHA). AIHA (AIHA), a disorder that affects 0.001% of the general population, involves autoantibody production against red cell antigens leading to shortened red cell survival [19]. The association of AIHA with bone marrow transplantation is well recognized. Allogenic bone marrow transplantation is a potential curative treatment for patients with acute leukemia. However, a substantial proportion of patients experience GVHD, as is the case with this patient. GVHD predisposes graft recipients to the development of AIHA. Evidence of hemolysis in this patient included high total bilirubin (primarily unconjugated), anemia, very high LD activity, and very low haptoglobin concentrations. A few cases of congenital absence of haptoglobin have been described especially among Blacks of West African origin (Nigeria, Cameroon) where there is a prevalence of approximately 30% [20]. Increased post-transplantation LD may be associated with host organ damage, repopulation of activated donor lymphocytes, and hemolysis. A recent study demonstrated that a high blood cyclosporine concentration and a low LD activity were the most important parameters associated with a lower incidence of GVHD [21]. The initial measurement of serum LD and serial follow-up testing has been proposed as a prognostic tool for the response to chemotherapy [22]. An elevated LD at diagnosis was found to be associated with an increased probability of acute myeloid leukemia (AML) evolution and decreased probability of survival [23]. Increased LD activity has been associated with decreased survival following allogenic bone marrow transplant in a heterogeneous population of patients in whom 30% had acute leukemia [24]. Data have also shown that high LD activity independently predicts poor prognosis following allogenic bone marrow transplant for AML [25]. It is expected that serum LD would decrease after splenectomy and this should serve as a good prognostic tool for recovery and an indicator of survival.

8.2 Biochemistry and physiology

8.2.1 Physiological function

LD (EC 1.1.1.27, L-Lactate:NAD+ oxidoreductase) is a cytoplasmic enzyme with a ubiquitous distribution in vertebrate organisms [26]. LD functions in the final step of glycolysis and plays an important role in tissues that utilize glucose such as the myocardium, erythrocytes, liver, and skeletal muscle. It is a hydrogen transfer enzyme that catalyzes the reduction of pyruvate to lactate using reduced nicotine adenine dinucleotide (NADH) as the hydrogen donor. It plays a vital role in glycolysis, permitting organisms to anaerobically reoxidize cytoplasmic NADH. It is also present in many invertebrates, plants, and microbes. In a few of these organisms, D-lactate acts as the substrate [27, 28].

LD is also capable of reducing other short chain α-keto carboxylic acids to their corresponding α-hydroxy acids [29]. The reaction is thermodynamically reversible and the equilibrium is pH dependent. Alkaline pH (9–10) favors formation of pyruvate, whereas pH in the range of 7–8 favors formation of lactate. In kinetic terms, reduction of pyruvate is far more rapid than oxidation of lactate. At physiological temperature and pH, the equilibrium constant is approximately 10^{10} in the direction of lactate formation [30].

8.2.2 Tissue sources

LD found in normal serum usually originates primarily from erythrocytes and platelets. High activities of the enzyme are found in serum upon release from cells and tissues. Mild tissue damage will give rise to significant serum LD activity because of the very high concentration in tissues compared with plasma (500 times more on average). Active LD is a tetramer with a molecular weight of 134,000 kDa and may be composed of three different polypeptides termed A, B, and C. The A and B polypeptides are also referred to as M (for muscle) and H (for heart), respectively, whereas the C polypeptide is expressed primarily in the testis. Each of these polypeptide products has a molecular weight of 34 kDa but differ in amino acid content, tertiary structure, charge, substrate specificity, and sensitivity to inhibitors, temperature, and electrophoretic mobility [31].

In mammals, the LD A, B, and C polypeptide chains are encoded by three separate genes. LD A and C genes are located on chromosome 11 and the LD B gene is on chromosome 12 [32]. Circulating LD consists almost exclusively of the H and M subunits leading to five tetrameric isoenzymes: LD_1 (HHHH; H_4), LD_2 (HHHM; H_3M_1), LD_3 (HHMM; H_2M_2), LD_4 (HMMM; H_1M_3), and LD_5 (MMMM; M_4). Several other isoenzymes have been described. LD X is a homotetramer of C subunits derived from the testes but is not found in appreciable amounts in serum. A reported sixth LD isoenzyme (LD_6) found in critically ill patients was later shown to be an alcohol dehydrogenase derived from liver, which also metabolizes lactate [33]. The presence of LD_{66}, or alcohol dehydrogenase, on electrophoresis is a poor prognostic sign because significant liver damage must be present before this enzyme is found.

The highest activities of LD are found in the brain, erythrocytes, kidneys, leukocytes, platelets, lungs, myocardium, skeletal muscle, liver, lymph nodes, and spleen [34]. The most anodal (at pH 8.6) LD isoenzymes (LD_1 and LD_2,) are predominant in the cardiac muscle, erythrocytes, and kidneys, whereas the most cathodal (LD_4 and LD_5) are predominant in the skeletal muscle and liver. LD_3, which has intermediate mobility, is frequently associated with other sources such as the lungs, spleen, platelets, leukocytes, and lymph nodes. In many cases, there is a general increase in all fractions, but there are still some conditions with specific isoenzyme increases such as: (i) elevation of LD_1 and LD_2 occurs in myocardial infarction, megaloblastic anemia, and renal infarction; (ii) LD_2 and LD_3 elevation occurs in acute leukemia; and (iii) LD_4 and LD_5 elevation occurs in hepatic and skeletal muscle damage.

LD isoenzyme measurements are rarely indicated today as there are more tissue-specific and sensitive markers for diagnosis of various diseases. LD is inactivated by endocytosis and lysosomal degradation in the liver [35, 36]. Small amounts of LD are present in normal urine and can only be measured after the urine is concentrated, although this assay is not useful for clinical purposes.

8.2.3 Reference ranges

Values for LD activity in serum vary considerably, depending on the direction of the enzyme reaction and the method used. The reference interval in adult Caucasian individuals, determined with an assay traceable to the IFCC reference procedure at 37°C, was found to be 125–220 U/L [37]. In a recent reference interval study of children 0–17 years of age that employed the IFCC recommended method, children 0–1 year (196–438 U/L) had the highest LD activities [38]. LD decreased gradually with age with the oldest age group (13–17 years of age) characterized by the interval 115–287 U/L. Studies done on cord blood demonstrated high neonatal activities (90–320 U/L) using the same IFCC traceable method [39].

Using the agarose gel technique with fluorometric quantitation of the generated NADH, the following reference intervals (expressed as a percentage of total LD) for isoenzymes was obtained [40]: LD_1 14–26%, LD_2 29–39%, LD_3 20–26%, LD_4 8–16%, LD_5 6–16%.

8.3 Chemical pathology

LD is a non-specific but sensitive indicator of tissue damage to major organs in the body. Several disorders have been associated with increases in LD and, because of its non-specificity, it is rarely used for diagnostic purposes but it is a more useful disease monitoring tool. LD is released into peripheral blood in cases of excessive cold or heat, dehydration, starvation, injury, and exposure to toxins (bacterial and chemical) and drugs. For clinical purposes, LD is useful in the field of cardiology, hepatology, hematology, and oncology [41].

8.3.1 Cardiology

The diagnostic utility of LD in cardiology has been reduced considerably with the emergence of more specific and sensitive markers such as troponin I and T. LD was used as a late marker of acute myocardial infarction (AMI). Concentrations of LD begin to rise 36 h post-infarction, peak at 48–72 h, and remain elevated for up to 12 days. The earliest sign of an AMI is the reversal of the ratio of LD_1:LD_2 (usually <1.0) referred to as the "LD flip" in which the ratio is >1.0. Current protocols for the diagnosis and monitoring of AMI do not include measurement of LD.

8.3.2 Hepatology

The concentration of LD in liver cells is approximately one-third of that in myocardial cells, but 1,800 times that in serum [42]. The determination of LD in liver disease lacks the specificity of other enzymes such as the transaminases and ALP. LD has been reported to be of diagnostic utility in toxic hepatitis, hepatic neoplasm, and liver transplantation. Varying sensitivities (52% to 100%) have been reported for LD as a marker for liver metastasis [43]. Measurement of LD isoenzymes has been used to increase the specificity compared with measurement of total enzyme activity. For example, a relative increase in LD_4 and LD_5 has been shown to detect metastatic liver cancer better than an elevation in total LD [43]. LD has not been useful in differentiating primary liver cell carcinoma from a secondary tumor. In orthotopic liver transplantation, the LD_5:LD_2 ratio seems to be a good indicator of early graft function and complications [44].

8.3.3 Hematology

High LD has been useful as a prognostic tool in acute and chronic lymphocytic leukemia. Acute lymphocytic leukemia is associated with the highest circulating LD activity. No appreciable increase in LD has been found in acute and chronic myeloid leukemia. Megaloblastic anemia has been associated with high activities of LD and thus LD may be used as a prognostic tool in these patients [45]. In megaloblastic anemia secondary to folate or B_{12} deficiency, there is ineffective erythropoiesis and hemolysis with release of large amounts of LD_1 and LD_2 causing a great increase in total LD activity. Owing to the very high levels of LD in erythrocytes, LD is a good marker of hemolysis and is useful in conditions such as sickle cell disease. LD has also been used as a prognostic marker of shock.

8.3.4 Oncology

LD is used for predicting survival in Hodgkin's lymphoma and NHL. A negative correlation was demonstrated between pretreatment serum LD and the probability of survival in these patients [46]. LD was also found to correlate well with disease progression in NHL with high-grade NHL having higher enzyme concentrations compared with low-grade NHL. In bone marrow transplant patients, high LD activity is associated with a 2.5-fold increase in relapse [47]. Total LD and LD isoenzyme activities have been used to predict the length of survival, response to chemotherapy, evaluation of tumor burden, and prognosis in patients with Burkitt's lymphoma [48]. Serum LD is thought to directly originate from the neoplastic cells in these lymphomas. Increased LD_1 has been reported in patients with germ cell tumors such as testicular seminomas, teratomas, and ovarian dysgerminomas, and has been reported to be of use in the staging, prognosis, and monitoring of patients with testicular germ cell tumors [49]. Elevated LD activity predicts inferior overall survival in all myeloma subgroups of the International Staging System and is of prognostic use in bone malignancies such as osteosarcoma [50–52].

LD has been used for a long time and is still useful in differentiating transudates from exudates in pleural and ascitic fluids. Light et al. demonstrated that a fluid to serum LD ratio >0.6 and a total fluid LD >200 U/L generally resulted from exudative effusions [53]. Such criteria used by Light et al. have been modified by various authors [54–56]. The measurement of LD and cholesterol in peritoneal fluid proved to be helpful in discriminating patients with ovarian cancer from those with benign ovarian tumor, particularly in the presence of negative cytology, increasing the diagnostic accuracy to 96% [57]. Peritoneal fluid LD has a positive correlation with advanced stage, poor histological type, higher grade, and positive abdominal cytology in ovarian cancer [58].

8.3.5 Neurology

There have been several reports on the usefulness of LD and LD isoenzymes in cerebrospinal fluid in a variety of neurological disorders. LD isoenzymes seem to play discriminatory roles in diseases states. Vázquez et al. used LD to differentiate structural and non-structural neurological disorders, whereas Nussinovitch et al. found that bacterial meningitis has a low LD_2 compared with aseptic meningitis, which is associated with high LD_3 [59, 60]. LD has been reported to be of prognostic value in disorders such as multiple sclerosis, febrile convulsions, Guillain-Barre syndrome, Creutzfeldt-Jakob disease, and hydrocephalus [61–64].

8.3.6 Macro-LD

There have been reports of healthy patients with very high LD attributed to macroenzyme phenomena [65]. Three forms of macro-LD are known: (i) a complex of LD and immunoglobulin [IgA (60% of all cases), IgG, or IgM]; (ii) LD and β-lipoprotein; and (iii) self-association of the individual isoenzymes. Macro-LD may appear in electrophoresis as an abnormal number of bands, a change in the electrophoretic movement of the isoenzymes, or widening of an existing band [66–68]. The presence of very high LD in a patient may present a diagnostic dilemma and awareness of its existence may spare the patient from undergoing a needlessly exhaustive clinical investigation.

8.3.7 Genetic deficiencies

No individual has been described with absence of both LD M and H genes. Partial deficiency states of the M gene or H gene have been described [69, 70]. Extensive studies have been performed in Japan where the frequency of heterozygotes for either LD M or LD H deficiency was found to be 0.104% at each locus [71]. LD deficiency is of interest to laboratory medicine mainly because it can result in misdiagnosis in those disorders where an increased serum LD is expected. A persistently normal LD concentration can be used to rule out disorders of major organs. However, this can be misleading in genetic deficiency of either the M or H chains. LD M deficiency has been associated with myoglobinuria and risk of renal failure after strenuous exercise, whereas LD H deficiency has no clear symptomatic consequences.

8.4 Analysis

Current methods for quantitation of total LD activity employ kinetic spectrophotometry to measure the interconversion of the NAD^+ and NADH at 340 nm. The lactate to pyruvate reaction is more commonly utilized than the pyruvate to lactate reaction. The lactate to pyruvate reaction is less dependent on the concentration of cofactor NAD^+ (vs. NADH) and substrate (lactate). The pyruvate to lactate reaction uses smaller substrate concentrations, but has faster reaction kinetics leading to early loss of linearity [72, 73]. The optimum pH for the lactate to pyruvate reaction is 8.8–9.8, whereas that of pyruvate to lactate is 7.4–7.8. This pH varies with the predominant isoenzyme in the mixture and also depends on the temperature and buffer/substrate concentrations. The specificity of human LD extends from L-lactate to other related 2-hydroxy and oxo-acids, and does not act on D-lactate. Only NAD^+ serves as co-enzyme. Both pyruvate and lactate in excess inhibit enzyme activity with inhibition by either substrate being greater for the H than for the M chain. Substrate inhibition decreases with increasing pH. The catalytic oxidation of 2-hydroxybutyrate to 2-oxo-butyrate (acetoacetate) is accomplished by 2-hydroxybutyrate dehydrogenase but in a limited manner by LD. LD_1 and LD_2 oxidize 2-hydroxybutyrate more rapidly than LD_5. This property has been exploited in determining the LD_1 isoenzyme activity because most of the activity of 2-hydroxybutyrate dehydrogenase is due to the action of LD_1. The analytical goals for desirable performance of LD assays, derived from biological variation of the enzyme (~8.6%), are an analytical coefficient of variation of ≤4.3% and a total error of ±11.4% [74].

8.4.1 LD isoenzyme determination

Electrophoretic separation on agarose gels or cellulose acetate membranes in an alkaline buffer (pH 8.6) is followed by detection of activity in the gel. After the isoenzymes have been separated by electrophoresis (LD_1 being the most anodal and LD_5 the most cathodal), a reaction mixture containing L-lactate and NAD^+ is layered over the separation medium. The NADH generated over the LD zones is detected either by its fluorescence when excited by long-wave ultraviolet light or by its reduction of a tetrazolium salt to form a colored formazan. LD_1 can be assayed alone after inhibition of LD_2–LD_5 by chemical inhibitors such as sodium perchlorate and guanidine thiocyanate. LD_1 can also be measured by immunoprecipitation; LD_2–LD_5 are precipitated out of solution by an antisera leaving LD_1 to be assayed spectrophotometrically. The half-life of the LD isoenzymes varies from 4 to 4.5 days for LD_1 to 4 to 6 h for LD_5. The biological variability of the LD isoenzymes is between 4.8% and 12.4% [51].

8.4.2 IFCC reference method

A lactate to pyruvate reference method, optimized for LD_1 has been developed by the IFCC [75]. This method is based on the IFCC primary reference procedure for LD at 37°C [72]. The reaction of lactate and NAD^+ is allowed to occur at an alkaline pH at 37°C. The rate of formation of NADH at 340 nm is monitored. The results are expressed in U/L (quantity of LD that catalyzes the conversion of 1 μmol of lactate per minute). This can be converted to SI units (μkat/L) by multiplying by 0.017. The term "katal" refers to the number of moles of LD required to convert L-lactate to pyruvate per second.

8.4.3 Specimen

Serum is the preferred specimen for measuring LD activity and it is generally higher than plasma LD due to LD release from platelets during clotting [76]. Plasma samples may be contaminated with platelets, which contain high concentrations of LD. Serum should be separated from the clot as soon as possible after the specimen has been obtained. Hemolysis of any degree affects LD levels because erythrocytes contain 150 times more LD activity than serum [77, 78]. Transport through pneumatic tube systems can induce hemolysis resulting in mild LD increases.

LD isoenzymes vary in their sensitivity to cold, with LD_4 and LD_5 being especially labile. Activity of LD_4 and LD_5 is lost if the samples are stored at −20°C. Thus, serum specimens should be stored at room temperature or 4°C, at which no loss of activity occurs for at least 3 days. Transient increase in LD occurs after blood transfusion and also in patients with thrombocytosis (due to release from platelets). LD usually returns to normal in 24 h.

8.5 Questions and answers

1. LD found in normal serum is derived primarily from which of the following?
   (a) Erythrocytes and platelets
   (b) Liver
   (c) Skeletal muscle
   (d) Myocardium
   (e) Kidney
2. The enzymatic function of LD is to catalyze which of the following?
   (a) Phosphorylation of lactate during glycolysis
   (b) Reduction of pyruvate to lactate to regenerate NAD
   (c) Reduction of pyruvate to lactate to regenerate NADH
   (d) Reduction of lactate of pyruvate to regenerate NAD
   (e) Reduction of pyruvate to lactate to regenerate NADH
3. Which of the following disorders is associated with the highest serum LD activities?
   (a) Acute myocardial infarction
   (b) Renal infarction
   (c) Acute lymphocytic leukemia
   (d) Viral hepatitis
   (e) Alcoholic hepatitis
4. Which of the following disorders is typically associated with an increase in the LD_1 isoenzyme?
   (a) Skeletal muscle injury
   (b) Hepatitis

    (c) Pulmonary infarction
    (d) Orthotopic liver transplantation
    (e) Testicular seminoma

5.   Which of the following would likely be seen in an exudative effusion?
    (a) No measurable LD activity
    (b) A fluid-to-serum LD ratio of 0.6 or greater
    (c) Fluid LD activity that is less than 200 µ/L
    (d) Fluid LD activity that is at least two-fold greater than serum creatine kinase activity
    (e) Fluid LD activity that is at least two-fold greater than serum glucose concentrations

6.   Which of the following statements is correct?
    (a) Serum LD activity is higher than plasma LD due to inhibition of LD by heparin
    (b) Serum LD activity is lower than plasma LD due to uptake of LD by erythrocytes during clotting
    (c) Serum LD activity is higher than plasma LD due to release of LD from platelets during clotting
    (d) Serum LD activity is lower than plasma due to degradation of LD by platelets activated during the process of clotting
    (e) Serum LD activity is lower than plasma due to heparin-induced release of LD from erythrocytes

References

[1] Kim HH, Joo H, Kim T, Kim E, Park SJ, Park JK, et al. The mitochondrial Warburg effect, a cancer enigma. Interdisciplinary Bio Central 2010,1,1–6.

[2] Buccheri GF, Violante B, Sartoris AM, Ferrigno D, Curcio A, Vola F. Clinical value of a multiple biomarker assay in patients with bronchogenic carcinoma. Cancer 1986,57,2389–96.

[3] Rotenberg Z, Weinberger I, Sagie A, Fuchs J, Davidson E, Sperling O, et al. Total lactate dehydrogenase and its isoenzymes in serum of patients with non-small-cell lung cancer. Clin Chem 1988,34,668–70.

[4] Tanaka T, Fujii M, Nishikawa A, Bunai Y, Obayashi F, Sugie S, et al. A cytochemical study of lactic dehydrogenase (LDH) isoenzymes in human lung cancer. Cancer Detect Prev 1984,7,65–71.

[5] Turna A, Solak O, Cetinkaya E, Kiliçgün A, Metin M, Sayar A, et al. Lactate dehydrodgenase levels predict pulmonary morbidity after lung resection for non-small cell lung cancer. Eur J Cardiothorac Surg 2004,26,483–7.

[6] Karagiannis A, Tsorlalis I, Kakafika A, Pateinakis P, Perifanis V, Harsoulis F. Acute renal failure due to tumor lysis syndrome in a patient with non-Hodgkin's lymphoma. Ann Hematol 2005, 84,343–6.

[7] Kim S, Chung DH. Pediatric solid malignancies. Neuroblastoma and Wilms' tumor. Surg Clin North Am 2006,86,469–87.

[8] Pandian SS, McClinton S, Bissett D, Ewen SW. Lactate dehydrogenase as a tumour marker in adult Wilms' tumour. Br J Urol 1997,80,670–1.

[9] Chuang CK, Ng KF, Liao SK. Adult Wilms' tumor presenting as acute abdomen with elevated serum lactate dehydrogenase-4 and -5 isoenzymes. Case report. Chang Gung Med J 2000,23,235–9.

[10] den Boer ME, Wanders RJ, Morris AA, IJlst L, Heymans HS, Wijburg FA, et al. Long chain 3-hydroxyacyl-CoA dehydrogenase deficiency. Clinical presentation and follow-up of 50 patients. Pediatrics 2002,109,99–104.

[11] Hupuczi P, Nagy B, Sziller I, Rigó B, Hruby E, Papp Z. Characteristic laboratory changes in pregnancies complicated by HELLP syndrome. Hypertens Pregnancy 2007,26,389–401.

[12] Carpani G, Bozzo M, Ferrazzi E, D' Amato B, Pizzotti D, Radaelli T, et al. The evaluation of maternal parameters at diagnosis may predict HELLP syndrome severity. J Matern Fetal Neonatal Med 2003,13,147–51.

[13] Federico M, Molica S, Bellei M, Luminari S. Prognostic factors in low-grade non-Hodgkin lymphomas. Curr Hematol Malig Rep 2009,4,202–10.

[14] Karadeniz C, Oguz A, Citak EC, Uluoglu O, Okur V, Demirci S, et al. Clinical characteristics and treatment results of pediatric B-cell non-Hodgkin lymphoma patients in a single center. Pediatr Hematol Oncol 2007,24, 417–30.

[15] Ansell SM, Armitage J. Non-Hodgkin lymphoma: diagnosis and treatment. Mayo Clin Proc 2005,80,1087–97.

[16] Solal-Céligny P, Roy P, Colombat P, White J, Armitage JO, Arranz-Saez R, et al. Follicular Lymphoma International Prognostic Index. Blood 2004,104,1258–65.

[17] Federico M, Bellei M, Marcheselli L, Luminari S, Lopez-Guillermo A, Vitolo U. Follicular Lymphoma International Prognostic Index 2. A new prognostic index for follicular lymphoma developed by the International Follicular Lymphoma Prognostic Factor Project. J Clin Oncol 2009,27,4555–62.

[18] Bouafia F, Drai J, Bienvenu J, Thieblemont C, Espinouse D, Salles G, et al. Profiles and prognostic values of serum LD isoenzymes in patients with haematopoietic malignancies. Bull Cancer 2004,91,E229–40.

[19] Hashimoto C. Autoimmune hemolytic anemia. Clin Rev Allergy Immunol 1998,16,285–95.

[20] Allison AC, Blumberg BS, Rees W. Haptoglobin types in British, Spanish Basque and Nigerian African population. Nature 1958,181,824.

[21] Song MK, Chung JS, Seol YM, Kwon BR, Shin HJ, Choi YJ, et al. Influence of lactate dehydrogenase and cyclosporine A level on the incidence of acute graft-versus-host disease after allogeneic stem cell transplantation. J Korean Med Sci 2009,24,555–60.

[22] Hafiz MG, Mannan MA. Serum lactate dehydrogenase level in childhood acute lymphoblastic leukemia. Bangladesh Med Res Counc Bull 2007,33,88–91.

[23] Wimazal F, Sperr WR, Kundi M, Meidlinger P, Fonatsch C, Jordan JH, et al. Prognostic significance of serial determinations of lactate dehydrogenase (LDH) in the follow-up of patients with myelodysplastic syndromes. Ann Oncol 2008,19,970–6.

[24] Mehta J, Gordon LI, Tallman MA, Winter JN, Evens AM, Frankfort O, et al. Does younger age affect the outcome of reduced-intensity allogeneic hematopoietic stem cell transplantation for hematologic malignancies beneficially? Bone Marrow Transplant 2006,38,95–100.

[25] Kalaycio M, Rybicki L, Pohlman B, Dean R, Sweetenham J, Andresen S, et al. Elevated lactate dehydrogenase is an adverse predictor of outcome in HLA-matched sibling bone marrow transplant for acute myelogenous leukemia. Bone Marrow Transplant 2007,40, 753–8.

[26] Markert CL. Lactate dehydrogenase. Biochemistry and function of lactate dehydrogenase. Cell Biochem Funct 1984,2,131–4.

[27] Ellington WR, Long GL. Purification and characterization of a highly unusual tetrameric D-lactate dehydrogenase from the muscle of the giant barnacle *Balanus nubilus* Darwin. Archiv Biochem Biophys 1978,186,265–74.

[28] Long GL. D-Lactate dehydrogenase from the horseshoe crab. Methods Enzymol 1975,41,313–8.

[29] McComb RB. The measurement of lactate dehydrogenase. In: Homburger HA, ed. Clinical and Analytical Concepts in Enzymology. Skokie, IL, USA, College of American Pathologists, 1983,157–71.

[30] Menon MP, Hunter FR, Miller S. Kinetic studies on human lactate dehydrogenase isoenzyme catalysed lactate to pyruvate reaction. J Prot Chem 1987,6,413–29.

[31] Panteghini M, Bais R, van Solinge W. Enzymes. In: Burtis CA, Ashwood ER, Bruns DE, eds. Tietz Textbook of Clinical Chemistry and Molecular Diagnostics, 4th ed. St Louis, MO, USA, Elsevier Saunders, 2006,601–3.

[32] Li SS, Luedemann M, Sharief FS, Takano T, Deaven LL. Mapping of human lactate dehydrogenase-A, -B, and -C genes and their related sequences: the gene for LDHC is located with that for LDHA on chromosome 11. Cytogenet Cell Genet 1988,48,16–8.

[33] Kato S, Ishii H, Kano S, Horii K, Tsuchiya M. Evidence that "lactate dehydrogenase isoenzyme 6" is in fact alcohol dehydrogenase. Clin Chem 1984,30,1585–6.

[34] Dito WR. A simple time-saving method for interpretative report generation. I. Lactate acid dehydrogenase isozymes. Am J Clin Pathol 1973,59,439–47.

[35] Smit MJ, Beekhuis H, Duursma AM, Bouma JM, Gruber M. Catabolism of circulating enzymes: plasma clearance, endocytosis, and breakdown of lactate dehydrogenase-1 in rabbits. Clin Chem 1988,34,2475–80.

[36] Wilkinson JH, Qureshi AR. Catabolism of plasma enzymes, as studied with ^{125}I-labeled lactate dehydrogenase-1 in the rabbit. Clin Chem 1976,22,1269–76.

[37] Pagani F, Bonora R, Panteghini M. Reference interval for lactate dehydrogenase catalytic activity in serum measured according to the new IFCC recommendations. Clin Chem Lab Med 2003,41,970–1.

[38] Heiduk M, Päge I, Kliem C, Abicht K, Klein G. Pediatric reference intervals determined in ambulatory and hospitalized children and juveniles. Clin Chim Acta 2009,406,156–61.

[39] Freer DE, Statland BE, Johnson M, Felton H. Reference values for selected enzyme activities and protein concentrations in serum and plasma derived from cord-blood specimens. Clin Chem 1979,25,565–9.

[40] McKenzie D, Henderson RA. Electrophoresis of lactate dehydrogense isoenzymes. Clin Chem 1983,29,189–95.

[41] Huijgen HJ, Sanders GT, Koster RW, Vreeken J, Bossuyt PM. The clinical value of lactate dehydrogenase in serum: a quantitative review. Eur J Clin Chem Clin Biochem 1997,35,569–79.

[42] Lott JA, Nemesanzky E. Lactate dehydrogenase (LD). In: Lott LA, Wolf PL, eds. Clinical Enzymology. New York, NY, USA, Field, Rich, and Associates, 1986,213–44.

[43] Rotenberg Z, Welnberge I, Davidson E, Fuchs J, Harell D, Agmon J. Lactate dehydrogenase isoenzyme patterns in serum of patients with metastatic liver disease. Clin Chem 1989,35,871–3.

[44] Rodrigue F, Boyer O, Feillet F, Lemonnier A. Lactate dehydrogenase isoenzyme LD5/LD2 ratio as an indicator of early graft function and complications following pediatric orthotopic liver transplantation. Transplant Proc 1995,27,1871–4.

[45] Jaswal TS, Mehta HC, Gupta V, Singh M, Singh S. Serum lactate dehydrogenase in diagnosis of megaloblastic anaemia. Indian J Pathol Microbiol 2000,43,325–9.

[46] Lindh J, Lenner P, Osterman B, Roos G. Prognostic significance of serum lactic dehydrogenase levels and fraction of S-phase cells in non-Hodgkin lymphomas. Eur J Haematol 1993,50, 258–63.

[47] Conde E, Sierra J, Iriondo A, Domingo A, García Laraña J, Marín J, et al. Prognostic factors in patients who received autologous bone marrow transplantation for non-Hodgkin's lymphoma. Report of 104 patients from the Spanish Cooperative Group GEL/TAMO. Bone Marrow Transplant 1994,14,279–86.

[48] Zinzani PL, Gherlinzoni F, Bendandi M, Salvucci M, Tura S. Adult Burkitt's lymphoma: clinical and prognostic evaluation of 20 patients. Leuk Lymphoma 1994,14,465–70.
[49] von Eyben FE. A systematic review of lactate dehydrogenase isoenzyme 1 and germ cell tumors. Clin Biochem 2001,34,441–54.
[50] Terpos E, Katodritou E, Roussou M, Pouli A, Michalis E, Delimpasi S, et al. High serum lactate dehydrogenase adds prognostic value to the international myeloma staging system even in the era of novel agents. Eur J Haematol 2010,85,114–9.
[51] Anagnostopoulos A, Gika D, Symeonidis A, Zervas K, Pouli A, Repoussis P, et al. Multiple myeloma in elderly patients: prognostic factors and outcome. Eur J Haematol 2005,75,370–5.
[52] Bacci G, Longhi A, Ferrari S, Briccoli A, Donati D, De Paolis M, et al. Prognostic significance of serum lactate dehydrogenase in osteosarcoma of the extremity. Experience at Rizzoli on 1421 patients treated over the last 30 years. Tumori 2004,90,478–84.
[53] Light RW, MacGregor MI, Luchsinger PC, WC Ball, et al. Pleural effusions: the diagnostic separation of transudates and exudates. Ann Intern Med 1972,77,508–13.
[54] Paramothayan NS, Barron J. New criteria for the differentiation between transudates and exudates. J Clin Pathol 2002,55,69–71.
[55] Joseph J, Badrinath P, Basran GS, Sahn SA. Is the pleural fluid transudate or exudate? A revisit of the diagnostic criteria. Thorax 2001,56,867–70.
[56] Maranhão BH, Silva CT, Chibante AM, Cardoso GP. Determination of total proteins and lactate dehydrogenase for the diagnosis of pleural transudates and exudates: redefining the classical criterion with a new statistical approach. J Bras Pneumol 2010,36,468–74.
[57] Halperin R, Hadas E, Bukovsky I, Schneider D. Peritoneal fluid analysis in the differentiation of ovarian cancer and benign ovarian tumor. Eur J Gynaecol Oncol 1999,20,40–4.
[58] Yüce K, Baykal C, Genç C, Al A, Ayhan A. Diagnostic and prognostic value of serum and peritoneal fluid lactate dehydrogenase in epithelial ovarian cancer. Eur J Gynaecol Oncol 2001,22,228–32.
[59] Vázquez JA, Adducci Mdel C, Monzón DG, Iserson KV. Lactic dehydrogenase in cerebrospinal fluid may differentiate between structural and non-structural central nervous system lesions in patients with diminished levels of consciousness. J Emerg Med 2009,37,93–7.
[60] Nussinovitch M, Finkelstein Y, Elishkevitz KP, Volovitz B, Harel D, Klinger G, et al. Cerebrospinal fluid lactate dehydrogenase isoenzymes in children with bacterial and aseptic meningitis. Transl Res 2009,154,214–8.
[61] Nussinovitch M, Prais D, Finkelstein Y, Harel D, Amir J, Volovitz B. Lactic dehydrogenase isoenzymes in cerebrospinal fluid of children with Guillain-Barré syndrome. Arch Dis Child 2002,87,255–6.
[62] Nussinovitch M, Volovitz B, Finkelstein Y, Amir J, Harel D. Lactic dehydrogenase isoenzymes in cerebrospinal fluid associated with hydrocephalus. Acta Paediatr 2001,90,972–4.
[63] Schmidt H, Otto M, Niedmann P, Cepek L, Schröter A, Kretzschmar HA, et al. CSF lactate dehydrogenase activity in patients with Creutzfeldt-Jakob disease exceeds that in other dementias. Dement Geriatr Cogn Disord 2004,17,204–6.
[64] Nussinovitch M, Avitzur Y, Finkelstein Y, Amir J, Harel D, Volovitz B. Lactic dehydrogenase isoenzyme in cerebrospinal fluid of children with febrile convulsions. Acta Paediatr 2003,92,186–9.
[65] Perry C, Peretz H, Ben-Tal O, Eldor A. Highly elevated lactate dehydrogenase level in a healthy individual. A case of macro-LDH. Am J Hematol 1997,55,39–40.
[66] Podlasek SJ, McPherson RA, Threatte GA. Specificity of autoantibodies to lactate dehydrogenase isoenzyme subunits. Clin Chem 1985,31,527–32.
[67] Biewenga J, Feltkamp TEW. Lactate dehydrogenase (LDH)-IgG3 immunoglobulin complexes in human serum. Clin Chim Acta 1975,64,101–16.

[68] Peters O, Gorus K, Van Camp B. NAD+ dissociable macromolecular lactate dehydrogenase. Clin Chem 1982,28,1826–7.

[69] Kanno T, Sudo K, Takeuchi I, Kanda S, Honda N, Nishimura Y, et al. Hereditary deficiency of lactate dehydrogenase M-subunit. Clin Chim Acta 1980,108,267–76.

[70] Houki N, Matsushima Y, Kitamura M, Tukada T, Nishina T, Nakayama T. A case of deficiency of lactate dehydrogenase H-subunit. Jpn J Clin Chem 1986,15,85–90.

[71] Maekawa M, Sudo K, Nagura K, Li SS, Kanno T. Population screening of lactate dehydrogenase deficiencies in Fukuoka Prefecture in Japan and molecular characterization of three independent mutations in the lactate dehydrogenase-B(H) gene. Hum Genet 1994,93,74–6.

[72] Schumann G, Bonora R, Ceriotti F, Clerc-Renaud P, Ferrero CA, Férard G, et al. IFCC Primary reference procedures for the measurement of catalytic activity concentrations of enzymes at 37°C. Part 3. Reference procedure for the catalytic measurement of lactate dehydrogenase. Clin Chem Lab Med 2002,40,643–8.

[73] Howell BF, McClure S, Schaffer R. Lactate-to-pyruvate or pyruvate-to-lactate assay for lactate dehydrogenase: re-examination. Clin Chem 1979,25,269–72.

[74] Ricós C, Alvarez V, Cava F, García-Lario JV, Hernández A, Jiménez CV, et al. Current databases on biologic variation: pros, cons and progress. Scand J Clin Lab Invest 1999,59,491–500.

[75] Bais, R, Philcox M. Approved recommendation of IFCC methods for the measurement of catalytic concentration of enzymes. Part 8. IFCC method for lactate dehydrogense (L-Lactate:NAD+ oxidoreductase, EC 1.1.1.27). International Federation of Clinical Chemists (IFCC). Eur J Clin Chem Clin Biochem 1994,32,639–55.

[76] Bais R, Edwards JB. Plasma lactate dehydrogenase activity will be increased if detergent and platelets are present. Clin Chem 1977,23,1056–8.

[77] Frank JJ, Bermes EW, Bickel MS, Watkins BF. Effect of in vitro hemolysis on chemical values for serum. Clin Chem 1978,224,1966–70.

[78] Yucel D, Dalva K. Effect of in vitro hemolysis on 25 common biochemical tests. Clin Chem 1992,38,575–7.

9 Pancreatic lipase

Wan-Ming Zhang, Edmunds Reineks, Joe M. El-Khoury
and Sihe Wang

9.1 Case studies

9.1.1 Patient A

A 68-year-old female presented with severe cramping with intermittent stabbing pain. Her symptoms started 4 months ago. Her pain was in the epigastric area with radiation to her back. She had a past medical history of hypertension, asthma, and arthritis. The patient denied fever or chills, but had significant nausea. She claimed having a regular diet and denied foul smelling or loose stools. She was initially hospitalized for 2 weeks until her serum enzymes decreased and she tolerated food intake by mouth. She was readmitted 1 week later with recurrence of abdominal pain and elevation of pancreatic enzymes. An initial computed tomography (CT) of the abdomen showed development of a pancreatic pseudocyst, 6 cm × 4 cm. A subsequent CT of the abdomen showed enlargement of the pseudocyst, 10 cm × 6 cm. She denied alcohol and smoking history. She had no medications prior to the recent episode, and was intermittently on an angiotensin-converting-enzyme inhibitor. There was no evidence of hypercalcemia or autoimmune pancreatitis.

On physical examination, the patient's abdomen was mildly distended with diffuse tenderness to palpation. This tenderness was worse in the epigastric and left flank areas. The patient felt abdominal tenderness with percussion and with voluntary guarding. There was no rebound tenderness. Laboratory tests on admission were: white blood cell (WBC) count 16,100/μL, red blood cell (RBC) count 3.27×10^6/μL, hemoglobin 90 g/L, hematocrit 29.4%, platelet count 500,000/μL, prothrombin time (PT) 35 s, partial thromboplastin time (PTT) 11 s, blood urea nitrogen 10.7 mmol/L, aspartate transaminase (AST) 27 U/L, alanine transaminase (ALT) 20 U/L, amylase 820 U/L, and lipase 1,073 U/L.

9.1.1.1 Discussion

This patient was diagnosed with acute pancreatitis and a hemorrhagic pancreatic pseudocyst. The most common causes for acute pancreatitis are cholelithiasis and alcohol abuse [1]. Pancreatitis can also arise as a side effect of medication use, from iatrogenic causes, or from other etiologies [1, 2]. Acute pancreatitis is characterized by acute inflammation and necrosis of pancreatic parenchyma, focal enzymatic necrosis of peri-pancreatic fat, and vessel necrosis, which may cause hemorrhage. These pathological changes are due to intra-pancreatic activation of pancreatic enzymes and enzymatic digestion of pancreatic and peri-pancreatic tissue. Lipase activation produces necrosis of adipose tissue in the pancreatic interstitial and peri-pancreatic spaces. Calcium precipitates can form in the necrotic tissue. Digestion of vascular walls results in thrombosis and hemorrhage. The inflammation and necrosis inside the pancreatic tissue cause significant leakage of pancreatic enzymes into the circulation that appear as elevations of serum pancreatic lipase, amylase, and other pancreatic enzymes. Pancreatic triglyceride lipase (PTL) is usually elevated within 4–8 h after onset of acute pancreatitis and remains elevated for 7–14 days [3]. Acute pancreatitis also results in extravasation of pancreatic enzymes, which can digest the adjacent tissues. This results in a collection of fluid around the pancreas that contains pancreatic enzymes, hemolyzed blood, and necrotic tissue debris. These collections may resolve spontaneously if they have a small volume. However, those with larger volumes may evolve into a pancreatic pseudocyst, with a thick wall of granulation tissue and fibrosis [4]. A small pseudocyst that causes no symptoms may be managed conservatively. However, a large pseudocyst will need intervention such as endoscopic trans-gastric drainage. This patient had a progressive enlargement of her pseudocyst, as evidenced by CT imaging. Her low hemoglobin and hematocrit may indicate active bleeding inside the pseudocyst. After stabilization of her acute episode, the patient underwent endoscopic retrograde cholangiopancreatography (ERCP)-aided drainage which produced 1.9 L of dark fluid, containing blood. The patient was given antibiotics for possible pseudocyst infection. She tolerated food by mouth well and was discharged home in a stable condition 3 days later.

9.1.2 Patient B

A 45-year-old male patient with a history of chronic pancreatitis secondary to alcohol abuse presented with a 3-day history of uncontrollable central abdominal pain, which radiated to his back. His pain was relieved by sitting forward and is exacerbated by the smell of food or eating food. He reported that the pain was 10/10, associated with nausea and vomiting. He denied any change of bowel habits and he reported no diarrhea. He stated that his appetite had decreased dramatically secondary to his nausea and vomiting. The patient had a history of crack cocaine and tobacco abuse. His last alcoholic beverage was a week ago. The patient recalled that the first similar episode of abdominal pain was 4 years ago, and he had since had three episodes due to alcohol abuse. He required hospitalization for each of these occurrences.

On physical examination, the patient was cachectic with considerable weight loss and skeletal muscle atrophy. He had tenderness at the right upper quadrant and epigastric areas. There was evidence of hepatomegaly. There was no flank pain. Bowel sounds were present. There was no rigidity, guarding, or rebound.

A complete blood count (CBC) analysis showed WBC 15,700 cells/μL, hemoglobin 150 g/L, hematocrit 44%, and platelets 226,000/μL. Serum tests showed triglycerides 2.23 mmol/L, urea nitrogen 2.14 mmol/L, creatinine 70.7 μmol/L, calcium 2.06 mmol/L,

AST 21 U/L, ALT 28 g/L, alkaline phosphatase (ALP) 114 U/L, total bilirubin 6.8 µmol/L, albumin 28 g/L, total protein 65 g/L, lipase 852 U/L, and amylase 305 U/L. CT imaging of the abdomen showed that the patient had a dilated common bile duct, which measured 11.8 mm, along with liver changes consistent with fatty infiltration. He was noted to have dilatation of his pancreatic duct measuring 7.7 mm in caliber along with calcification seen at the head of the pancreas.

9.1.2.1 Discussion
This patient was diagnosed with acute onset of chronic pancreatitis. Excessive alcohol use is the most common cause of chronic pancreatitis and can also be a cause of acute pancreatitis [1, 2]. The diagnosis of chronic pancreatitis is typically based on tests examining pancreatic structure and function. Repeat attacks of chronic pancreatitis cause a gradual loss of pancreatic exocrine function that may result in persistent diarrhea, increased fecal fat, and lack of elevation in serum values of amylase and lipase. A secretin stimulation test is considered the gold standard test for diagnosis of pancreatic function, but is not often used clinically [5]. Other common tests used to evaluate chronic pancreatitis are fecal elastase measurement, serum trypsinogen, a CT scan, ultrasound, endoscopic ultrasound, magnetic resonance imaging, ERCP, and magnetic resonance cholangiopancreatography (MRCP). Pancreatic calcification can often be seen on plain abdominal X-rays, as well as on CT scans. The patient was hospitalized and started on antibiotics secondary to his leukocytosis. His initial abdominal CT showed some pancreatic necrosis. He was given intravenous fluid support and opiates for pain. After his abdominal pain was completely resolved, his diet was advanced from liquids to soft food at discharge. A repeat CT showed improvement of his pancreatitis.

9.1.3 Patient C

A 64-year-old male came to the hospital for evaluation of persistent diarrhea. He complained that he had a lengthy history of diarrhea which dated back to approximately 8 months ago. He described the stool as being loose, water-like, and lighter in color, but not with any obvious steatorrhea or blood. His diarrhea occurred approximately 1–2 h after eating meals, especially after fatty foods. He denied any history of nausea, vomiting, or low-grade fevers. He also denied severe pain in the abdomen, but did complain of ongoing weight loss of approximately 18.1 kg (40 lbs) over the past 7–8 months. He was evaluated for his diarrhea with no significant improvement. His family history was significant for coronary artery disease, breast cancer, and diabetes.

Physical examination revealed that there was no peripheral lymphadenopathy, and his abdomen was soft and non-tender. There was no evidence of any masses or organomegaly. Bowel sounds were present. There was no pruritus or icterus.

Laboratory results were: fasting blood glucose 6.94 mmol/L, CA-19.9 191 KU/L, carcinoembryonic antigen 14.3 µg/L, lipase 256 U/L, and amylase 233 U/L. His stool cultures and assay of Clostridium difficile toxin were negative. Evaluation of his diarrhea included a colonoscopy but the results were non-contributory. Further evaluation included a CT scan of his abdomen, which showed a 2.5-cm pancreatic mass with possible encasement of the superior mesenteric artery. Also noted was a 1.2-cm periportal lymph node as well as several 2-mm nodules in the right and left lower lung fields. An endoscopic ultrasound was performed with an attempt to do a transduodenal fine needle aspiration (FNA). Unfortunately, the FNA was not successful due to encasement of the superior mesenteric artery.

9.1.3.1 Discussion

Ultimately, the patient's diagnosis was pancreatic carcinoma with involvement of local lymph nodes. The encasement of the superior mesenteric artery excluded the possibility of resection. The carcinoma was located in the head of pancreas. It is likely that the tumor mass caused blockage of exocrine secretion, increasing the duct pressure. The higher secretion pressure caused more enzyme back-diffusion into the circulation, resulting in increased amylase and lipase activities in serum [6]. However, the tumor probably did not involve the biliary tract, because no icterus was found. In the advanced stages of pancreatic cancer, these enzymes may decrease due to extensive destruction of pancreatic tissue [7]. Reduction of pancreatic enzymes significantly affects digestion and absorption of food nutrients, especially proteins and fats, as in this patient, who initially complained of significant diarrhea and weight loss. The patient underwent several rounds of chemotherapy. He died of metastatic pancreatic cancer 2 years later.

9.1.4 Patient D

A 13-year-old male, accompanied by his mother, presented for continued management of cystic fibrosis (CF). According to his mother, the patient initially presented with pancreatitis at the age of 5 years, and was soon diagnosed with CF. He had recurrent pancreatitis, which resulted in several hospitalizations. The patient had at least two additional episodes of recurrent pancreatitis, which were managed at home. He had always been tall and skinny, and there had been some concern regarding his ability to gain weight. He was started on supplemental pancreatic enzymes at the age of 6 years and had gained approximately 6.8 kg (15 lbs) since then. There was no concern with his liver function. The patient had routine bowel movements at least twice a day. He had had constipation in the past, which was managed with medication. An MRCP was performed to evaluate his pancreatic ducts, but the study was inconclusive. He had an ERCP, which showed a normal caliber pancreatic duct without strictures or filling defects. He frequently experienced abdominal pain due to acute onset of chronic pancreatitis after eating. He did not typically vomit when he had the pancreatitis, but felt nauseated. The pain intensity was 2/10.

Laboratory results for serum showed amylase 352 U/L, lipase 475 U/L, albumin 46 g/L, total bilirubin 6.84 μmol/L, conjugated bilirubin 1.71 μmol/L, ALP 203 U/L, AST 24 U/L, ALT 11 U/L, vitamin A 1.22 μmol/L, 1,25-$(OH)_2$-vitamin D 138.24 pmol/L, and total protein 71 g/L. His trypsin in fecal samples was significantly lower than normal, whereas sweat chloride was abnormally increased.

9.1.4.1 Discussion

The diagnosis for this patient was chronic pancreatitis due to CF. CF causes failure in the synthesis of the functional protein, CF transmembrane conductance regulator (CFTR), which is important in cross-membrane uptake of sodium and chloride, resulting in dehydrated and thick mucus in the lungs, pancreas, and intestines [8]. These thickened secretions in CF can cause narrowing in the passages of affected organs, leading to remodeling and infection in the lungs, damage by accumulated digestive enzymes such as lipase in the pancreas, and blockage of the intestines by thick feces. The patient frequently experienced respiratory infections, abdominal pain due to pancreatitis and constipation, and malnutrition due to pancreatic enzyme insufficiency. During acute episodes of pancreatitis, his serum pancreatic enzymes were elevated. Frequent pancreatitis causes extensive damage to pancreatic function, resulting in exocrine enzyme insufficiency. His symptoms were managed by daily use of pancreatic enzyme preparations and multiple fat-soluble vitamin supplements, antibiotics for prevention of pulmonary or other mucosal infections, and inhalation therapy to decrease the thickness of viscous purulent secretions. He was presently asymptomatic, but he experienced fevers almost every month in conjunction with respiratory or digestive symptoms. Repeat infections and progressive damage to affected organs in CF significantly affect the life expectancy of CF patients. It is estimated that life expectancy is 37.4 years for newborns with CF who were born in or after 2008 [9]. The patient was regularly followed for his serum pancreatic enzymes, and given protease tablets as supplements for his pancreatic insufficiency.

9.2 Biochemistry and physiology

9.2.1 Molecular forms

Lipase (EC 3.1.1.3) is a water-soluble hydrolase that catalyzes the hydrolysis of ester bonds in triglycerides. It belongs to a subclass of esterases. However, lipase specifically acts as an ester-water interface to cleave emulsified long-chain triglycerides at a higher rate than any other esterase [10, 11]. Human serum lipases are mainly derived from pancreatic tissue, where they are synthesized and stored in granules [12]. Human pancreatic lipases are composed of human triglyceride pancreatic lipase (PTL) and pancreatic lipase-related proteins 1 and 2 (PLRP1 and PLRP2). The predominant pancreatic lipase is PTL with a mean PLRP2 to PTL mass ratio of 28.3% [13]. The term "lipase" used in this chapter is specifically restricted to human PTL unless otherwise specified.

9.2.2 Molecular structure and tissue expression

PTL is encoded by the *PNLIP* gene (human pancreatic lipase gene). The *PNLIP* gene is located in gene locus 10q26.1 [14]. The *PNLIP* gene encodes the PTL protein with a single chain of 465 amino acids, in which the first 16 amino acids comprise the signal peptide. PTL has no pro-peptide, and activation of PTL relies on the conformational change in the presence of lipid substrates [15, 16]. Purified human PTL has a molecular mass of 48 kDa and contains a single N-oligosaccharide chain [17].

PTL contains two distinct domains. The N-terminal domain, containing residues 1 to 336, has a structure known as the α/β-hydrolase fold, which is also present in other lipases and esterases [16]. N-terminal domain residues Ser 153, Asp 177, and His 263 form the catalytic site. These residues are conserved in PTL, PLRP1, and PLRP2, and the structure around the catalytic serine has only been found in lipases and esterases [18, 19]. PTL contains a surface loop defined by a disulfide bridge between residues Cys 238 and Cys 262. The surface loop is also called the lid domain, which forms van der Waals interactions with the β5 loop of residues 76–85 and β9 of 204–224. The lid domain, together with the β5 and β9 loops, sterically hinders access of substrate to the active center, constituting an inactive state [20]. Structural studies predicted that the lid domain contributes to the substrate specificity and lipid-binding property of PTL, and this prediction was later supported by results from site-mutation studies [20]. PTL with a mutated lid domain had significantly lower activities towards triglyceride-derived substrates. The C-domain of PTL contains residues 337–465. It has a β-sandwich structure and provides the major binding surface for colipase [21]. Colipase is a co-factor for PTL and is exclusively produced by the pancreas, see Section 2.4.5. Residues 405–414 form a hydrophobic loop, which participates in lipid binding [22].

Expression of mRNA encoding PTL is mainly in the exocrine pancreas, although low levels of PTL can be detected in other tissues, such as extrahepatic peribiliary glands [23, 24]. In one study, the newborn rat pancreas expresses little or no mRNA encoding PTL [25]. Lipase levels in the human duodenum increase with age and reach adult levels in the first 1–2 years of life [26].

9.2.3 Mechanism of catalysis

9.2.3.1 Conformational activation
The lid domain in the N-terminal of PTL sterically hinders the enzymatic active center. PTL must undergo a conformational change before it can actively catalyze the hydrolysis of substrate. Structural studies of a crystallized PTL-colipase complex demonstrated the active conformation of PTL [27]. In the presence of micelles of lipid substrates, the lid domain, together with the other hydrophobic loops, β5 and β9, move away from the active center and present an open conformation [21, 28]. The enzymatic active center is exposed to the lipid substrates following movement of the lid domain. In addition, once movement of the lid domain takes place, several new interactions, such as hydrogen bonds, are formed between the residues in colipase and PTL. These newly formed interactions stabilize the active conformation state [28]. The ternary complexes, composed of PTL-colipase and bile salt micelles, exist in solution, as evidenced by neutron diffraction studies [29]. These complexes are driven to the emulsion surface by hydrophobic interactions.

9.2.3.2 Mechanism of catalysis
PTL employs a very similar hydrolysis mechanism as found in other serine proteases. The crystal structure reveals a Ser 153-His 264-Asp 177 triad in PTL, in which Ser 153 is located in the center of the catalytic site. Ser 153 is a nucleophilic residue, like that in serine proteases, which is essential for catalytic activity [18]. After binding of the ester substrate, a tetrahedral intermediate is formed by nucleophilic attack of Ser 153, with the oxyanion stabilized by two or three hydrogen bonds, the so-called "oxyanion hole". The ester bond is then cleaved and the alcohol moiety released [30]. In the last step, the acyl enzyme is hydrolyzed. The nucleophilic attack by the catalytic serine is mediated by His 264 and Asp 177. PTL can cleave esters of medium- to long-chain fatty acids, but hydrolyzes long-fatty acid esters (C16–C22) more efficiently than other lipases [10, 30].

9.2.4 Functional regulation

9.2.4.1 Effect of pH
The optimal pH for enzymatic activity of PTL *in vitro* ranges from 7.4 to 10, depending on the type of substrate and buffer. PTL is stable under a wide range of pH values from 6.5 to 8.0 in the small intestine, but its activity favors more neutral pH conditions [31]. The pH value in the intestinal lumen shows an increase from duodenum to distal ileum. The pH value in the small intestine lumen is regulated

by secretion of bicarbonate from hepatic and pancreatic ducts and intestinal epithe-lium [32]. The pH value in the intestinal lumen can be affected by various diseases. For instance, an intestinal pH of 5.0 was found in patients with CF [33]. Abnormal pH levels in the small intestine lumen resulting from disease processes may impact the lipolytic activity of PTL.

9.2.4.2 Lipase inhibitors

Orlistat and its derivatives are inhibitors of PTL. These inhibitors are saturated deri-vatives of lipstatin, a lipid of microbial origin [34, 35]. *Orlistat* specifically targets the substrate-binding site of PTL and is used as a drug for obesity. Other inhibitors for PTL include organophosphates [36], carbodiimides [37], and iodine [7].

9.2.4.3 Bile salts

Bile acids are produced in the liver and stored in the gallbladder. The majority of bile acids are conjugated to glycine and taurine within hepatocytes to increase their solu-bility. The conjugated bile salts and unconjugated bile acids act as biological deter-gents to emulsify dietary lipids, and make them suitable substrates for hydrolysis by PTL [32]. Bile salts efficiently inhibit lipoprotein lipase, which interferes with the measurement of PTL in serum, because lipoprotein lipase is also found in serum and hydrolyzes triglycerides [38]. Excess bile salts could also inhibit PTL because they can reduce the interfacial adhesion of PTL [39]. However, under physiological conditions, bile salts and bile acids increase the area of water-oil interface in the small intestine lumen through emulsification of triglyceride droplets, thereby increasing the lipase-substrate interaction [32]. Furthermore, the inhibition of PTL by bile salts can be res-tored by colipase [27, 40].

9.2.4.4 Calcium

The *in vitro* addition of calcium to the enzymatic reaction leads to a significant incre-ase in the lipolytic activity of purified PTL [41]. Calcium does not bind PTL or colipase, but it can increase the reaction velocity and increase the affinity of PTL for substrate. This observation led some investigators to speculate that calcium may reduce the lag phase of PTL-mediated hydrolysis [42]. When the concentration of calcium is increa-sed to more than 40 μM, the maximum reaction velocity appears to plateau and the substrate affinity increases. It is not known if calcium has any impact on intestinal PTL activity under physiological condition because the concentration of calcium nor-mally exceeds 40 μM. There are several mechanisms that may explain the impact of calcium on PTL activity. Calcium may have a protective effect on PTL activity because calcium can precipitate free fatty acids released from hydrolysis of triglycerides by

PTL, and these free fatty acids are toxic to the active center of PTL. Calcium can reduce the electrostatic repulsion of bile salts to lipase, and therefore increase the interaction between PTL and substrates [7]. Addition of calcium into the assay mixture is an important step in many clinical assays for PTL activity [41, 43, 44].

9.2.4.5 Colipase

Pancreatic colipase is a 12-kDa, single-chain protein that contains a signal peptide of 23 amino acids, a pro-peptide of 5 amino acids, and mature peptide of 112 amino acids [45]. Colipase is encoded by the colipase gene that is located at gene locus 6p21.1. It is produced primarily in pancreatic tissue and is secreted as a pro-form into the duodenum, where pro-colipase is activated by trypsin. The 5-amino acid pro-peptide released upon activation is called enterostatin, which is a hormone for regulating satiety and fat intake [46]. Colipase mRNA was mainly expressed in pancreatic tissue, but low levels of colipase mRNA have also been found in rat gastric epithelium [45, 47]. The major function of colipase is to anchor lipase non-covalently to the surface of lipid micelles and to restore the enzymatic activity of PTL that may be inhibited by intestinal bile salts [48]. A recent study also indicated that colipase stabilizes the lid domain of lipase in the open conformation, thereby facilitating lipolysis [27]. Deficiency of colipase has been reported in patients with long-term steatorrhea due to poor lipolysis [49].

9.2.5 Lipase in serum

Fractionation of serum from healthy human subjects by electrophoresis shows two bands with lipolytic activity in the α- and γ-globulin regions [50]. A similar finding was also observed in pancreatic juice. The molecular mass of these two forms of lipase was estimated to be 39 and >200 kDa [50]. In patients with pancreatitis, four bands have been detected with lipolytic activity [7]. However, it was not shown that these bands all contained PTL, because early measurements of lipolysis were not specific to PTL, and some forms were not reactive to monoclonal antibodies later raised against PTL [7, 51]. The components with molecular mass >200 kDa may form from self-polymerization of PTL or by association with other proteins. Recent studies have shown that PTL is complexed with immunoglobulin G or α-2 macroglobulin, and these complexes are present in serum from patients with normal PTL or hyperlipasemia [52, 53]. The complexed forms of PTL maintain lipolytic activity against triglyceride-based substrates in the presence of colipase and bile salts. The presence of the complexed forms of PTL in serum raises the possibility that these complexed forms of PTL may cause discrepancies in assay results when different methods are used [7].

9.2.6 Other lipases

9.2.6.1 Bile salt-dependent lipase

Bile salt-dependent lipase (BSDL) is predominantly expressed by the acinar cells of the pancreas and the lactating mammary gland [54]. It can hydrolyze triglycerides, phospholipids, cholesterol esters, and lipid-soluble vitamins [55]. The BSDL gene is located at gene locus 9q34.3 and encodes a peptide of 722 amino acids, corresponding to a molecular mass of approximately 80 kDa [56, 57]. The fully glycosylated protein is approximately 100–110 kDa and commonly appears in pancreatic juice and milk [58]. Because neonates have very low levels of PTL and colipase, the presence of high levels of BSDL in mammary milk suggests that BSDL acts as a substitute for PTL activity in promoting lipid digestion and absorption in the neonate. In cases of pancreatic disease such as pancreatitis, blood levels of BSDL are increased together with other pancreatic enzymes due to leakage into the circulation. Recent studies showed that plasma BSDL is associated with low-density lipoprotein (LDL), and oxidized LDL stimulated synthesis of plasma BSDL in macrophages and endothelial cells [59]. These findings suggest that BSDL may play a role in lipid metabolism and atherosclerosis [60].

9.2.6.2 Hepatic lipase

Hepatic lipase (HL) is expressed on the surface of hepatocytes and endothelial cells lining hepatic sinusoids, the adrenal glands, and the ovaries. HL is encoded by the *LIPC* gene that is located at gene locus 15q21–23 [61]. HL can hydrolyze triglycerides and phospholipids in all lipoproteins. However, the main function of HL is to convert intermediate density lipoproteins to LDL and the post-prandial, triglyceride-rich high-density lipoprotein (HDL) into the post-absorptive, triglyceride-poor HDL [62]. HL also plays a role as a ligand-bridging factor for receptor-mediated lipoprotein uptake [63]. HDL concentrations are inversely related to HL activity, and HL deficiency results in increased HDL cholesterol. Using an animal model, Karackattu et al. showed that HL deficiency led to a delay in the onset of atherosclerosis [64].

9.2.6.3 Lipoprotein lipase

Lipoprotein lipase (LPL; EC 3.1.1.34) is an enzyme that hydrolyzes lipids such as triglycerides in lipoproteins. It also functions as a ligand-bridging factor for receptor-mediated lipoprotein uptake into cells [65]. LPL is encoded by the *LPL* gene that is located at gene locus 8p22 [66]. LPL protein is present at the luminal surface of the capillary endothelium, but the mRNA is found exclusively in parenchymal cells, indicating that LPL is synthesized in parenchymal cells and then translocated to luminal surfaces [67]. The concentrations of serum LPL have significant relationships with serum lipids and lipoproteins, visceral fat area, insulin resistance, and the development of atherosclerosis [68, 69]. LPL has a paradoxical impact on the formation of atherosclerosis. When LPL is expressed by cells in the vessel wall, particularly macrophages, it is pro-atherogenic, whereas LPL produced by adipose and muscle tissue is anti-atherogenic [70, 71].

9.2.6.4 Hormone-sensitive lipase

Hormone-sensitive lipase (HSL) is an intracellular neutral lipase that is capable of hydrolyzing triacylglycerols, diacylglycerols, monoacylglycerols, and cholesteryl esters. It is encoded by the *LIPE* gene that is located at 19q13.1–q13.2 [72]. HSL is highly expressed in adipose tissue and steroidogenic tissues, with lower amounts expressed in cardiac and skeletal muscle, macrophages, and pancreatic islets [73, 74]. Expression of HSL is tightly regulated by certain hormones, such as insulin, through reversible phosphorylation and translocation [75, 76]. Physiologically, HSL is involved in adipocyte lipolysis, steroidogenesis, spermatogenesis, and the secretion and action of insulin. It may also play a role in pathological conditions, such as obesity, type 2 diabetes, and hyperlipidemia [76, 77].

9.2.6.5 Gastric lipase and lingual lipase

These lipases belong to the α/β-hydrolase fold superfamily. Each of these two lipases possesses a classical Asp-His-Ser catalytic triad as found in PTL, indicating structural and functional similarity [78]. Gastric lipase is secreted by gastric chief cells in the fundic mucosa of the stomach, whereas lingual lipase is produced by serous glands under the circumvallate and foliate papillae on the surface of the tongue [79, 80]. The optimal pH for hydrolytic activity ranges from pH 3 to 6 for gastric lipase, and 4.5 to 5.4 for lingual lipase. A key difference from PTL is that gastric and lingual lipases do not require bile acid or colipase for optimal enzymatic activity [81, 82]. Both gastric and lingual lipases catalyze the removal of one fatty acid from each triacylglycerol molecule, whereas PTL can simultaneously take two fatty acids, from positions 1 and 3, of each triacylglycerol. The diacylglycerol produced from hydrolysis by gastric and lingual lipases cannot be transported across the epithelium-lined gastrointestinal tract. Diacylglycerol can be further hydrolyzed into fatty acid and glycerol by PTL [83]. Therefore, both gastric and lingual lipases are less efficient compared with PTL, and they participate in early steps of fat digestion. Gastric and lingual lipases contribute approximately 30% of lipolytic activity during digestion in adults. Gastric and lingual lipases provide up to 50% of total lipolytic activity in neonates due to low levels of PTL in this group. In the case of decreasing PTL production associated with pancreatic dysfunction, both gastric and lingual lipase can partially compensate for loss of PTL in the digestion of lipids; for instance, lingual lipase contributes to 90% of fat digestion in patients with CF [84].

9.2.6.6 Pancreatic lipase-related proteins 1 and 2 (PLRP1 and PLRP2)

The *PNLIP* gene family is composed of three members, including *PTL*, *PLRP1*, and *PLRP2*. These three genes share extensive homology, indicating that these genes are derived from the same ancestor gene [85]. PLRP1 displays no significant activity towards any of the substrates tested, and its physiological role is still unknown. The sequence similarity between PLRP1 and *PTL* implies similar functional activity. For instance, both enzymes can catalyze the hydrolysis of triglycerides. However, there are clear differences among these two enzymes. *PLRP2* has the ability to hydrolyze galactolipids and phospholipids except triglycerides, but PTL only hydrolyzes triglycerides [86]. Human PLRP2 is found in pancreatic juice at birth, whereas PTL is not expressed at this stage of development. This finding suggests that PLRP2 may play an important role in dietary fat digestion of human milk [87]. Another difference is that human PLRP2 shows very weak interaction with colipase. Colipase cannot restore human PLRP2 activity that is inhibited by bile salts [88, 89]. Therefore, hydrolysis of triglyceride by human PLRP2 may play a role only in neonates, in which both PTL and bile acid are not abundant. The *PLRP2* gene disruption in transgenic mice supports the idea that PLRP2 may play a role of this kind just after birth and during the lactation period, because a decrease in the neonatal dietary fat absorption rates has been observed in suckling transgenic mice [90, 91].

9.2.7 Reference range and standardization

Newborns have lower serum levels of PTL compared with adults because of low expression of PTL in pancreatic tissue [25, 26]. The elderly (>70 years old) have higher levels of serum PTL compared with younger adults. Reference ranges are not comparable between different assays because it is difficult to standardize methods that have different assay mechanisms, kinetics, and substrates [7]. Even with the same assay format, differences in calibrators, incubation times, assay kinetics, and substrates can produce different results [7]. Each clinical laboratory should determine its own reference range. Assay specificity also affects reference ranges. Early assays measure PTL activity along with the activity of lipoprotein lipase or other lipases. The reference ranges determined by these early methods are higher than those obtained later by more specific assays.

9.3 Chemical pathology

9.3.1 Acute pancreatitis

Pancreatitis is an inflammatory disease of the pancreas. It presents in two different forms. Acute pancreatitis has a sudden onset, whereas chronic pancreatitis often presents with recurring or persistent abdominal pain with or without pancreatic exocrine deficiency. The most common etiologies for acute pancreatitis are excessive alcohol usage and gallstone disease, comprising 70–80% of total cases [1, 2]. Initiation of acute pancreatitis starts with unregulated activation of trypsin within pancreatic acinar cells. The activated enzymes in the pancreas lead to the autolysis of the gland and local inflammation [92]. Acute pancreatitis presents as an acute inflammation, and is followed by necrosis of pancreatic parenchyma, peri-pancreatic fat and vessels. These pathological changes are due to enzymatic digestion of pancreatic and peri-pancreatic tissue. The inflammation and necrosis inside the pancreatic tissue cause significant leakage of pancreatic enzymes into the circulation, resulting in an elevation of serum pancreatic enzymes, including lipase and amylase. The diagnosis of acute pancreatitis is often difficult because clinical presentations in the early stage are non-specific, often occurring as steady, dull tenderness in the epigastric region or left upper quadrant. Up to 30% of patients with acute pancreatitis have a morphologically normal pancreas by imaging studies such as CT or ultrasound [93]. The gold standard for diagnosis of pancreatitis is laparotomy and direct visualization of the pancreas. However, it is contradictory to perform laparotomy in suspected acute pancreatitis [94]. Several laboratory tests are used to aid in the diagnosis of acute pancreatitis. Measurements of amylase and PTL in serum are widely used for acute pancreatitis, whereas other serum protein tests such as trypsinogen, chymotrypsin, elastase, and ribonuclease are less frequently used because of relatively lower specificity and sensitivity [94, 95]. PTL starts to increase within 4–8 h after the onset of acute pancreatitis, peaks at 24 h, and returns to normal within 7–14 days [3]. Amylase shows a similar increase in the initial phase of acute pancreatitis as that of lipase, but it starts to decrease at 24 h [96]. Therefore, lipase in serum has a greater clinical sensitivity for acute pancreatitis after 24 h of onset compared with amylase.

9.3.2 Pancreatic cancer

Pancreatic cancer typically originates in the exocrine pancreas and more than 90% are adenocarcinomas [97, 98]. In the early stage, it often does not cause symptoms. As the disease progresses, patients may show symptoms such as back pain, weight loss, or jaundice. The prognosis for patients with pancreatic cancer is poor, with less than 5% survival 5 years after the diagnosis [99]. There are several tumor markers for detection of pancreatic cancer, including CA-19.9. However, the available markers lack clinical sensitivity and specificity, and none can be used alone for early detection [100]. However, when measurement of CA-19.9 was used in combination with imaging techniques, such as endoscopic ultrasound plus FNA, the clinical sensitivity can be as high as 90% [101]. Pancreatic cancer can have variable impact on the serum activity of PTL and other pancreatic enzymes. Depending on the stage and location, pancreatic cancer can cause a decrease in intestinal pancreatic exocrine enzymes, but increases in serum PTL. If the tumor is located at the head of the pancreas, it is likely that the tumor mass will cause blockage of exocrine secretion, increasing the duct pressure. The higher duct pressure can cause more enzyme back-diffusion into the circulation. Also, tumor mass may directly invade blood vessels, especially in advanced stages, causing greater leakage of enzymes into the blood. However, serum PTL may be decreased if most (>90%) of the normal pancreatic tissue is destroyed [102].

9.3.3 Cystic fibrosis

CF is caused by a mutation in the gene for the protein CFTR. The most common mutation is ΔF508, a deletion of three nucleotides that results in a loss of the amino acid phenylalanine at the 508th position on the protein [103, 104]. This mutation, as well as some others, can cause the CFTR protein to be less stable, increase its degradation, or decrease its production. The CFTR protein is anchored to the outer membrane of cells in sweat glands, lungs, pancreas, and other affected organs [105]. The protein spans the membrane and acts as a channel controlling ion movement, such as chloride, between the cytoplasm and the surrounding fluid. Children with CF present with acute pancreatitis at an early age and frequently experience recurrent acute-onset episodes. CF patients lose most of their pancreatic function by adult age, and most require supplementation with pancreatic enzyme preparations. In younger CF patients, the disease may cause increased serum PTL, but the older CF patients have lower serum activities of PTL, due to pancreatic insufficiency. It has been suggested that measurement of serum PTL could be used as a newborn screening test for suspected CF [106].

9.3.4 Diabetic ketoacidosis

Diabetic ketoacidosis (DKA) is a potentially life-threatening complication in patients with diabetes mellitus, especially in those with type 1 diabetes. Due to a lack of insulin, patients begin to metabolize fatty acids as opposed to glucose and produce ketone bodies. The typical clinical presentation of DKA includes nausea and vomiting, pronounced thirst, excessive urine production, severe abdominal pain, and "Kussmaul respiration" [107]. Adult patients with DKA may present with acute pancreatitis [108, 109]. For those patients with DKA but no radiographic evidence of pancreatitis, especially in children, the mechanisms that cause elevated serum pancreatic enzymes are not known. It has been suggested that hypoperfusion or ischemia may be factors in pancreatic injury in children with DKA [110]. Insidious injuries to the pancreatic acinar cells may lead to PTL or other pancreatic enzymes leaking into the circulation. Another explanation is that DKA causes moderate hypovolemia, which results in a reduced glomerular filtration rate. This may lead to a less efficient clearance of lipase or other pancreatic enzymes by the kidneys, resulting in increased serum activities of these enzymes in the circulation [111, 112]. However, extra-pancreatic production should also be considered as a potential or partial mechanism [113, 114]. The dysmetabolic state in DKA may trigger release of amylase from salivary glands [115]. Increases in non-pancreatic lipolytic enzymes released into the circulation may cause high serum PTL activity in patients with DKA, because some of the current tests used for measurement of PTL cannot differentiate PTL from non-pancreatic lipases [7, 114, 116].

9.3.5 Chronic renal failure

It has been reported that serum PTL and other pancreatic enzymes are increased in chronic renal failure (CRF) [117, 118]. A reduced glomerular filtration rate may be responsible for a mild increase in serum PTL, as well as increased serum amylase in patients with CRF, because both enzymes are removed by glomerular filtration [112]. It has not been determined why some CRF patients, especially those with long periods of hemodialysis, have extremely high serum activities of PTL and amylase [119]. Recent studies have suggested that increased release of pancreatic enzymes due to pancreatic inflammation might be a cause [117]. Alternatively, metabolic derangement and hemodynamic alterations that occur in uremic patients undergoing prolonged hemodialysis may provide another explanation [119]. It has been reported that the increased intestinal permeability in renal failure is linked to increased serum activities of amylase and lipase [120].

9.3.6 Iatrogenic effects

ERCP and endoscopic sphincterotomy are widely used to diagnose and treat biliary and pancreatic disease. After these procedures, most patients experience transient increases of serum PTL, normally within 48 h [121, 122]. The increase of serum PTL may be due to minor tissue injury during mechanical manipulation or a result of the contrast medium used [122]. Prolonged post-procedural increases in serum PTL may indicate more severe injury during the procedure, which may cause post-ERCP pancreatitis (PEP). Most cases of PEP are of mild to moderate severity and require hospitalization for 24–48 h.

9.3.7 Drug effects

Many therapeutic drugs can cause acute pancreatitis and increased serum PTL. The World Health Organization database lists a total of 525 drugs from various substance classes that can possibly induce acute pancreatitis as a side effect [123]. The incidence of drug-induced pancreatitis is rare, and the overall incidence is approximately 0.1–2% of all cases [124]. The incidence may be significantly increased in certain populations, which includes pediatric or elderly patients, females, immunocom-promised patients, as well as patients with inflammatory bowel disease [125–127]. Because most of the information related to drug-induced pancreatitis comes from single case or case series reports, and there are only a limited number of animal studies, the mechanisms behind most drug-induced pancreatitis cases are unknown. Drugs have been categorized as "definite", "probable", or "possible" in association with pancreatitis [128]. The diagnosis of drug-induced pancreatitis is often difficult to make because there are no unique clinical, biochemical, or imaging features that can be used to differentiate drug-induced versus other types of pancreatitis. To render a diagnosis of drug-induced pancreatitis, it is important to confirm the diagnosis of pancreatitis, which in most cases would involve demonstrating increased serum pan-creatic enzymes such as PTL or amylase. The next is to exclude other common etiolo-gies of pancreatitis such as alcohol abuse, biliary tract disease, or gallstone cholecys-titis. Although some drugs have demonstrated risk for inducing pancreatitis, it is not recommended to use serum pancreatic enzyme tests to routinely monitor asymptoma-tic patients during the course of drug therapy. There are some drugs known to cause increased serum PTL without inducing pancreatitis [129]. It has been reported that intravenous injection of heparin can cause increased serum PTL activity. However, heparin actually increases lipoprotein lipase activity in serum, and this activity may be measured as PTL activity by certain lipase assays [130].

9.4 Analysis

9.4.1 Specimen

Serum is a suitable specimen for PTL analysis. PTL is stable in serum for more than 1 week at room temperature, 3 weeks at 4°C, and years at −20°C. Most assays for PTL that are used in clinical laboratories measure the lipolytic activity of PTL. Colipase is added to optimize the enzymatic reaction [131]. Calcium-binding anticoagulants, such as EDTA, citrate, and oxalate, should be avoided in specimens for PTL measure-ment because calcium is required to stabilize lipase activity [132]. PTL can be detected at very low levels in urine from normal individuals [133]. Some investigators evaluated pancreatic function by analysis of urine for lipolytic products, after oral administra-tion of iodine-131-labeled triglyceride analogs. The levels of lipolytic products in urine were highly correlated with intestinal PTL activity, which is indicative of pancreatic exocrine function [134].

9.4.2 Methods and instrumentation

Serum PTL can be determined by measurement of the specific lipolytic activity of PTL or by measuring the molecular mass of PTL. Methods for measuring lipolytic activity typically measure accumulation of products from hydrolysis, such as increase of fatty acid and NADH. Alternatively, they may measure the decreasing turbidity of substrates [7]. Currently available assays for PTL show more specificity for PTL activity than previous methods [116]. This is due to the use of more specific substrates for PTL, such as triglyceride-based substrates with long-fatty acid chains. Addition of bile salts, colipase and calcium into the assay mixture can increase the assay specificity for PTL activity because these additives can significantly eliminate lipolytic activity from other lipases such as lipoprotein lipase [131].

9.4.2.1 pH method

This method measures free fatty acids released from hydrolysis of triglyceride-based substrates by PTL. The measurement of free fatty acids is achieved by continuous titration with addition of alkaline solution [135]. Alternatively, fatty acids can be titrated using potentiometric or conductometric methods because the released fatty acid results in a change of potential or current, which can be correlated with lipolytic activity of PTL. With the conductometric method, the assay sensitivity was increased by using substrates with short-chain fatty acids, but the specificity for PTL was decreased [136]. Optimization of the pH method was performed by Tietz et al. by varying the amounts of colipase, calcium, and bile salts used in the reaction [131]. These optimized methods showed good correlations with other methods, such as turbidimetric or immunometric methods [131, 137]. pH methods are not currently used in typical clinical laboratories because of complexity and time constraints, but they may be used as reference methods [7].

9.4.2.2 Turbidimetric methods

The turbidimetric assay for PTL is based on measurement of a change in turbidity after substrate hydrolysis. The level of PTL activity is correlated with decreased absorbance due to hydrolysis of an emulsified substrate. The early turbidimetric assays for PTL used olive oil or triolein as substrates [138]. The PTL activity in serum measured by early tests was significantly contaminated by other lipases such as lipoprotein lipase [131, 139]. Assay specificity could be improved by addition of colipase, calcium, and bile salts into the assay mixture [131, 139, 140]. Hemolyzed samples, or samples containing high levels of rheumatoid factor (>5.5 g/L), may cause interference in turbidimetric assays [141].

9.4.2.3 Colorimetric and fluorometric methods

The early colorimetric assay for PTL used dimercaprol tributyrate (BALB) as the substrate [142]. The SH group of dimercaprol that is released by PTL-mediated hydrolysis is detected by use of a chromogenic reagent, 5,5′-dithiobis(2-nitrobenzoic acid) (DTNB). Product formation is measured at 412 nm. The BALB-DTNB assay has been fully automated and showed excellent correlation with a radiometric assay for PTL, and it also had increased analytical sensitivity [143]. Kitamura et al. assessed the pancreatic insufficiency of various chronic pancreatic diseases and found that measurement of PTL with the BALB-DTNB method was better than measuring amylase to assess pancreatic response with cerulean and secretin stimulation methods [144]. Because the specificity for the determination of PTL was largely dependent on the presence of bile salts and colipase [145], it was unknown if the BALB-DTNB method also measured lipolytic activity from other lipases. By using a triglyceride-derived fluorescent substrate, pyrenedecanoyl-2,3-dioleyl glycerol (1P10TG), Salvayre et al. developed a fluorometric assay for PTL in which the fluorescently labeled fatty acid from hydrolysis was released into the aqueous phase where it could be measured [146]. The assay was optimized by adjusting the pH to 9.0 and included co-factors such as bile salts, calcium, and colipase in the mixture [146, 147]. The fluorometric assay did not measure lipolytic activity from neutral-alkaline leukocytic lipase. The included cofactors in the assay mixture improved the linearity of the kinetics, shortened the lag phase, and enhanced the activity by attaching the enzyme onto the oil microemulsion and by protecting the lipase from denaturation. Fossati et al. developed a colorimetric assay for PTL in which the substrate was a natural, micellar-solubilized long-chain fatty acid 1,2-diglyceride (mainly with palmitic and oleic acids) obtained from egg lecithin [148]. The substrate with long-chain fatty acids eliminated non-specific lipolytic activity from lipases other than PTL. The quantification of PTL activity was achieved through a glycerol kinase-coupled enzyme system to produce a quinine monoamine dye that can be measured by absorbance at 550 nm. The clinical significance of this assay was validated in a study that showed the specificity for diagnosis of acute pancreatitis reached 91% at a cut-off value of 57 U/L, with sensitivity of 100% [149]. Free fatty acid released from the lipolytic reaction of PTL was further broken down by β-oxidation through a coupled enzyme system to produce NADH. NADH production was measured spectrophotometrically at 340 nm and was correlated with the lipolytic activity of PTL. PTL activity was also measured on a dry slide in which 1,2-dilineoylglycerol was used as a substrate [150].

9.4.2.4 Immunochemical methods

Immunochemical methods offer an alternative for analysis of lipase in serum or other body fluids. Most immunoassays for lipase use dual antibodies specific to different epitopes of PTL. One of the two specific antibodies is coated onto the solid phase (capture antibody) and the other is conjugated with a signal molecule (detector). The immunoassays for PTL are not interfered by extra-pancreatic lipases because the antibodies used are highly specific and are shown to only react with PTL [151]. Immunoassays for PTL had excellent correlation with enzymatic assay methods, such as the turbidimetric assay [152]. By using a latex-agglutination test and a cut-off value of 300 μg/L, Møller-Petersen et al. found a diagnostic efficiency of 0.99 for acute pancreatitis [153]. In another study using a radioimmunoassay, Roberts and Mercer found the assay was 91% sensitive and 96% specific for the diagnosis of acute pancreatitis with a serum PTL of 112 ng/mL as the cut-off value [154]. Immunoassay methods tend to have higher sensitivity compared with enzymatic assays; therefore, they may be suitable for monitoring patients with pancreatic insufficiency.

9.5 Questions and answers

1. Human serum lipases are mainly derived from which tissue?
    (a) Renal
    (b) Duodenum
    (c) Pancreas
    (d) Hepatic
    (e) Ileum
2. Lipase requires which as a co-factor?
    (a) Zinc
    (b) Chloride
    (c) Vitamin B6
    (d) Colipase
    (e) Vitamin B12
3. What structural feature of lipase provides it with high selectivity towards hydrolyzing triglycerides?
    (a) The β-sandwich structure of the C-domain
    (b) The α/β-hydrolase fold of the N-domain
    (c) The N-terminal residues Ser 153, Asp 177, and His 263 that form the catalytic site
    (d) The lid domain
    (e) All of the above
4. All of the following lipases exist except?
    (a) Bile salt-dependent lipases
    (b) Hepatic lipase
    (c) Lipoprotein lipase
    (d) Hormone-sensitive lipase
    (e) Adrenal lipase
5. Measurement of serum lipase is helpful in all of the following conditions except?
    (a) Acute pancreatitis
    (b) Liver failure
    (c) Pancreatic cancer
    (d) Cystic fibrosis
    (e) Renal failure

6. Which of the following specimens is unacceptable for the measurement of serum lipase activity?
   (a) Serum separator
   (b) EDTA plasma
   (c) Plain serum
   (d) Sodium heparin
   (e) Lithium heparin
7. What methods have not been developed for the measurement of serum lipase activity?
   (a) Turbidimetric
   (b) pH-stat
   (c) Colorimetric and fluorometric
   (d) Immunochemical
   (e) High-performance liquid chromatography

References

[1] Yadav D, Hawes RH, Brand RE, Anderson MA, Money ME, Banks PA, et al. Alcohol consumption, cigarette smoking, and the risk of recurrent acute and chronic pancreatitis. Arch Intern Med 2009,169,1035–45.
[2] Johnson CD, Hosking S. National statistics for diet, alcohol consumption, and chronic pancreatitis in England and Wales, 1960–88. Gut 1991,32,1401–5.
[3] Matull WR, Pereira SP, O'Donohue JW. Biochemical markers of acute pancreatitis. J Clin Pathol 2006,59,340–4.
[4] Kalb B, Sarmiento JM, Kooby DA, Adsay NV, Martin DR. MR imaging of cystic lesions of the pancreas. Radiographics 2009,29,1749–65.
[5] Gullo L, Pezzilli R, Ventrucci M, Barbara L. Caerulein induced plasma amino acid decrease. A simple, sensitive, and specific test of pancreatic function. Gut 1990,31,926–9.
[6] Bakkevold KE, Arnesjo B, Kambestad B. Carcinoma of the pancreas and papilla of Vater: presenting symptoms, signs, and diagnosis related to stage and tumour site. A prospective multicentre trial in 472 patients. Norwegian Pancreatic Cancer Trial. Scand J Gastroenterol 1992,27,317–25.
[7] Tietz NW, Shuey DF. Lipase in serum – the elusive enzyme: an overview. Clin Chem 1993,39,746–56.
[8] Gadsby DC, Vergani P, Csanady L. The ABC protein turned chloride channel whose failure causes cystic fibrosis. Nature 2006,440,477–83.
[9] Jackson AD, Daly L, Kelleher C, Marshall BC, Quinton HB, Foley L, et al. The application of current lifetable methods to compare cystic fibrosis median survival internationally is limited. J Cyst Fibros 2011,10,62–5.
[10] Carriere F, Rogalska E, Cudrey C, Ferrato F, Laugier R, Verger R. In vivo and in vitro studies on the stereoselective hydrolysis of tri- and diglycerides by gastric and pancreatic lipases. Bioorg Med Chem 1997,5,429–35.
[11] Ferrato F, Carriere F, Sarda L, Verger R. A critical reevaluation of the phenomenon of interfacial activation. Methods Enzymol 1997,286,327–47.
[12] Yadav D, Agarwal N, Pitchumoni CS. A critical evaluation of laboratory tests in acute pancreatitis. Am J Gastroenterol 2002,97,1309–18.

[13] Eydoux C, Aloulou A, De Caro J, Grandval P, Laugier R, Carrière F, et al. Human pancreatic lipase-related protein 2. tissular localization along the digestive tract and quantification in pancreatic juice using a specific ELISA. Biochim Biophys Acta 2006,1760,1497–504.

[14] Davis RC, Diep A, Hunziker W, Klisak I, Mohandas T, Schotz MC, et al. Assignment of human pancreatic lipase gene (PNLIP) to chromosome 10q24–q26. Genomics 1991,11,1164–6.

[15] Lowe ME, Rosenblum JL, Strauss AW. Cloning and characterization of human pancreatic lipase cDNA. J Biol Chem 1989,264,20042–8.

[16] Ollis DL, Cheah E, Cygler M, Dijkstra B, Frolow F, Franken SM, et al. The alpha/beta hydrolase fold. Protein Eng 1992,5,197–211.

[17] De Caro A , Figarella C, Amic J, Michel R, Guy O. Human pancreatic lipase: a glycoprotein. Biochim Biophys Acta 1977,490,411–9.

[18] Lowe ME. The catalytic site residues and interfacial binding of human pancreatic lipase. J Biol Chem 1992,267,17069–73.

[19] Winkler FK, D'Arcy A, Hunziker W. Structure of human pancreatic lipase. Nature 1990,343,771–4.

[20] Yang Y, Lowe ME. The open lid mediates pancreatic lipase function. J Lipid Res 2000,41,48–57.

[21] van Tilbeurgh H, Sarda L, Verger R, Cambillau C. Structure of the pancreatic lipase-procolipase complex. Nature 1992,359,159–62.

[22] D'Silva S, Xiao X, Lowe ME. A polymorphism in the gene encoding procolipase produces a colipase, Arg92Cys, with decreased function against long-chain triglycerides. J Lipid Res 2007,48,2478–84.

[23] Terada T, Kida T, Nakanuma Y. Extrahepatic peribiliary glands express α-amylase isozymes, trypsin and pancreatic lipase: an immunohistochemical analysis. Hepatology 1993,18,803–8.

[24] Terada T, Nakanuma Y. Pancreatic lipase is a useful phenotypic marker of intrahepatic large and septal bile ducts, peribiliary glands, and their malignant counterparts. Mod Pathol 1993,6, 419–26.

[25] Payne RM, Sims HF, Jennens ML, Lowe ME. Rat pancreatic lipase and two related proteins: enzymatic properties and mRNA expression during development. Am J Physiol 1994,266, G914–21.

[26] Lebenthal E, Lee PC. Development of functional responses in human exocrine pancreas. Pediatrics 1980,66,556–60.

[27] Lowe ME. Structure and function of pancreatic lipase and colipase. Annu Rev Nutr 1997,17, 141–58.

[28] van Tilbeurgh H, Gargouri Y, Dezan C, Egloff MP, Nésa MP, Ruganie N, et al. Crystallization of pancreatic procolipase and of its complex with pancreatic lipase. J Mol Biol 1993,229,552–4.

[29] Lowe ME. The triglyceride lipases of the pancreas. J Lipid Res 2002,43,2007–16.

[30] Pleiss J, Fischer M, Schmid RD. Anatomy of lipase binding sites: the scissile fatty acid binding site. Chem Phys Lipids 1998,93,67–80.

[31] Iizuka K, Higurashi H, Fujimoto J, Hayashi Y, Yamamoto K, Hiura H. Purification of human pancreatic lipase and the influence of bicarbonate on lipase activity. Ann Clin Biochem 1991,28,373–8.

[32] Brownlee IA, Forster DJ, Wilcox MD, Dettmar PW, Seal CJ, Pearson JP. Physiological parameters governing the action of pancreatic lipase. Nutr Res Rev 2010,23,146–54.

[33] Youngberg CA, Berardi RR, Howatt WF, Hyneck ML, Amidon GL, Meyer JH, et al. Comparison of gastrointestinal pH in cystic fibrosis and healthy subjects. Dig Dis Sci 1987,32,472–80.

[34] Hochuli E, Kupfer E, Maurer R, Meister W, Mercadal Y, Schmidt K. Lipstatin, an inhibitor of pancreatic lipase, produced by *Streptomyces toxytricini*. II: chemistry and structure elucidation. J Antibiot (Tokyo) 1987,40,1086–91.

[35] Weibel EK, Hadvary P, Hochuli E, Kupfer E, Lengsfeld H. Lipstatin, an inhibitor of pancreatic lipase, produced by *Streptomyces toxytricini*. I: producing organism, fermentation, isolation and biological activity. J Antibiot (Tokyo) 1987,40,1081–5.

[36] Moreau H, Moulin A, Gargouri Y, Noël JP, Verger R. Inactivation of gastric and pancreatic lipases by diethyl p-nitrophenyl phosphate. Biochemistry 1991,30,1037–41.

[37] Semeriva M, Desnuelle P. Pancreatic lipase and colipase: an example of heterogeneous biocatalysis. Adv Enzymol Relat Areas Mol Biol 1979,48,319–70.

[38] Kinnunen PK, Huttunen JK, Ehnholm C. Properties of purified bovine milk lipoprotein lipase. Biochim Biophys Acta 1976,450,342–51.

[39] Gargouri Y, Julien R, Bois AG, Verger R, Sarda L. Studies on the detergent inhibition of pancreatic lipase activity. J Lipid Res 1983,24,1336–42.

[40] van Tilbeurgh H, Bezzine S, Cambillau C, Verger R, Carrière F. Colipase: structure and interaction with pancreatic lipase. Biochim Biophys Acta 1999,1441,173–84.

[41] Benzonana G. On the role of calcium ions during the hydrolysis of insoluble triglycerides by pancreatic lipase in the presence of bile salts. Biochim Biophys Acta 1968,151,137–46.

[42] Alvarez FJ, Stella VJ. The role of calcium ions and bile salts on the pancreatic lipase-catalyzed hydrolysis of triglyceride emulsions stabilized with lecithin. Pharm Res 1989,6,449–57.

[43] Brown WJ, Belmonte AA, Melius P. Effects of divalent cations and sodium taurocholate on pancreatic lipase activity with gum arabic-emulsified tributyrylglycerol substrates. Biochim Biophys Acta 1977,486,313–21.

[44] Leger C, Charles M. Pancreatic lipase. World Rev Nutr Diet 1980,35,96–128.

[45] Lowe ME, Rosenblum JL, McEwen P, Strauss AW. Cloning and characterization of the human colipase cDNA. Biochemistry 1990,29,823–8.

[46] Erlanson-Albertsson C, York D. Enterostatin – a peptide regulating fat intake. Obes Res 1997,5,360–72.

[47] Winzell MS, Lowe ME, Erlanson-Albertsson C. Rat gastric procolipase: sequence, expression, and secretion during high-fat feeding. Gastroenterology 1998,115,1179–85.

[48] Dahim M, Brockman H. How colipase-fatty acid interactions mediate adsorption of pancreatic lipase to interfaces. Biochemistry 1998,37,8369–77.

[49] Gaskin KJ, Durie PR, Lee L, Hill R, Forstner GG. Colipase and lipase secretion in childhood-onset pancreatic insufficiency: delineation of patients with steatorrhea secondary to relative colipase deficiency. Gastroenterology 1984,86,1–7.

[50] Kimura H, Kitamura T, Tsuji M. Studies on human pancreatic lipase. I. Interconversion between low and high molecular forms of human pancreatic lipase. Biochim Biophys Acta 1972,270,307–16.

[51] Arzoglou PL, Lessinger JM, Ferard G. Plasma lipase properties as related to pancreatic condition. Clin Chem 1986,32,50–2.

[52] Keating JP, Lowe ME. Persistent hyperlipasemia caused by macrolipase in an adolescent. J Pediatr 2002,141,129–31.

[53] Taes YE, Louagie H, Yvergneaux JP, De Buyzere ML, De Puydt H, Delanghe JR, et al. Prolonged hyperlipasemia attributable to a novel type of macrolipase. Clin Chem 2000,46,2008–13.

[54] Baba T, Downs D, Jackson KW, Tang J, Wang CS. Structure of human milk bile salt activated lipase. Biochemistry 1991,30,500–10.

[55] Lombardo D, Deprez P, Guy O. Esterification of cholesterol and lipid-soluble vitamins by human pancreatic carboxyl ester hydrolase. Biochimie 1980,62,427–32.

[56] Reue K, Zambaux J, Wong H, Lee G, Leete TH, Ronk M, et al. cDNA cloning of carboxyl ester lipase from human pancreas reveals a unique proline-rich repeat unit. J Lipid Res 1991,32, 267–76.

[57] Taylor AK, Zambaux JL, Klisak I, Mohandas T, Sparkes RS, Schotz MC, et al. Carboxyl ester lipase: a highly polymorphic locus on human chromosome 9qter. Genomics 1991,10, 425–31.

[58] Blackberg L, Hernell O. The bile-salt-stimulated lipase in human milk: purification and characterization. Eur J Biochem 1981,116,221–5.

[59] Caillol N, Pasqualini E, Mas E, Guieu R, Valette A, Boyer J, et al. Pancreatic bile salt-dependent lipase activity in serum of normolipidemic patients. Lipids 1997,32,1147–53.

[60] Li F, Hui DY. Modified low density lipoprotein enhances the secretion of bile salt-stimulated cholesterol esterase by human monocyte-macrophages: species-specific difference in macrophage cholesteryl ester hydrolase. J Biol Chem 1997,272,28666–71.

[61] Ameis D, Stahnke G, Kobayashi J, McLean J, Lee G, Büscher M, et al. Isolation and characterization of the human hepatic lipase gene. J Biol Chem 1990,265,6552–5.

[62] Connelly PW, Hegele RA. Hepatic lipase deficiency. Crit Rev Clin Lab Sci 1998,35,547–72.

[63] Kounnas MZ, Chappell DA, Wong H, Argraves WS, Strickland DK. The cellular internalization and degradation of hepatic lipase is mediated by low density lipoprotein receptor-related protein and requires cell surface proteoglycans. J Biol Chem 1995,270,9307–12.

[64] Karackattu SL, Trigatti B, Krieger M. Hepatic lipase deficiency delays atherosclerosis, myocardial infarction, and cardiac dysfunction and extends lifespan in SR-BI/apolipoprotein E double knockout mice. Arterioscler Thromb Vasc Biol 2006,26,548–54.

[65] Nykjaer A, Nielsen M, Lookene A, Meyer N, Røigaard H, Etzerodt M, et al. A carboxyl-terminal fragment of lipoprotein lipase binds to the low density lipoprotein receptor-related protein and inhibits lipase-mediated uptake of lipoprotein in cells. J Biol Chem 1994,269,31747–55.

[66] Mattei MG, Etienne J, Chuat JC, Nguyen VC, Brault D, Bernheim A, et al. Assignment of the human lipoprotein lipase (LPL) gene to chromosome band 8p22. Cytogenet Cell Genet 1993,63,45–6.

[67] Braun JE, Severson DL. Regulation of the synthesis, processing and translocation of lipoprotein lipase. Biochem J 1992,287,337–47.

[68] Kobayashi J, Nakajima K, Nohara A, Kawashiri M, Yagi K, Inazu A, et al. The relationship of serum lipoprotein lipase mass with fasting serum apolipoprotein B-48 and remnant-like particle triglycerides in type 2 diabetic patients. Horm Metab Res 2007,39,612–6.

[69] Kobayashi J, Nohara A, Kawashiri MA, Inazu A, Koizumi J, Nakajima K, et al. Serum lipoprotein lipase mass: clinical significance of its measurement. Clin Chim Acta 2007,378,7–12.

[70] Mead JR, Cryer A, Ramji DP. Lipoprotein lipase, a key role in atherosclerosis? FEBS Lett 1999,462,1–6.

[71] Mead JR, Ramji DP. The pivotal role of lipoprotein lipase in atherosclerosis. Cardiovasc Res 2002,55,261–9.

[72] Levitt RC, Liu Z, Nouri N, Meyers DA, Brandriff B, Mohrenweiser HM. Mapping of the gene for hormone sensitive lipase (LIPE) to chromosome 19q13.1→q13.2. Cytogenet Cell Genet 1995,69,211–4.

[73] Kraemer FB, Patel S, Saedi MS, Sztalryd C. Detection of hormone-sensitive lipase in various tissues. I. Expression of an HSL/bacterial fusion protein and generation of anti-HSL antibodies. J Lipid Res 1993,34,663–71.

[74] Kraemer FB, Patel S, Singh-Bist A, Gholami SS, Saedi MS, Sztalryd C. Detection of hormone-sensitive lipase in various tissues. II: regulation in the rat testis by human chorionic gonadotropin. J Lipid Res 1993,34,609–16.

[75] Cook KG, Yeaman SJ, Strålfors P, Fredrikson G, Belfrage P. Direct evidence that cholesterol ester hydrolase from adrenal cortex is the same enzyme as hormone-sensitive lipase from adipose tissue. Eur J Biochem 1982,125,245–9.

[76] Garton AJ, Campbell DG, Cohen P, Yeaman SJ. Primary structure of the site on bovine hormone-sensitive lipase phosphorylated by cyclic AMP-dependent protein kinase. FEBS Lett 1988,229,68–72.

[77] Yeaman SJ. Hormone-sensitive lipase – new roles for an old enzyme. Biochem J 2004,379,11–22.

[78] Roussel A, Canaan S, Egloff MP, Rivière M, Dupuis L, Verger R, et al. Crystal structure of human gastric lipase and model of lysosomal acid lipase, two lipolytic enzymes of medical interest. J Biol Chem 1999,274,16995–7002.

[79] Basque JR, Menard D. Establishment of culture systems of human gastric epithelium for the study of pepsinogen and gastric lipase synthesis and secretion. Microsc Res Tech 2000,48,293–302.

[80] Hamosh M, Scow RO. Lingual lipase and its role in the digestion of dietary lipid. J Clin Invest 1973,52,88–95.

[81] DeNigris SJ, Hamosh M, Kasbekar DK, Fink CS, Lee TC, Hamosh P. Secretion of human gastric lipase from dispersed gastric glands. Biochim Biophys Acta 1985,836,67–72.

[82] Roberts IM. Rat lingual lipase: effect of proteases, bile, and pH on enzyme stability. Am J Physiol 1985,249,G496–500.

[83] Hamosh M. Lingual and gastric lipases. Nutrition 1990,6,421–8.

[84] Abrams CK, Hamosh M, Hubbard VS, Dutta SK, Hamosh P. Lingual lipase in cystic fibrosis. Quantitation of enzyme activity in the upper small intestine of patients with exocrine pancreatic insufficiency. J Clin Invest 1984,73,374–82.

[85] Giller T, Buchwald P, Blum-Kaelin D, Hunziker W. Two novel human pancreatic lipase related proteins, hPLRP1 and hPLRP2. Differences in colipase dependence and in lipase activity. J Biol Chem 1992,267,16509–16.

[86] Andersson L, Carriére F, Lowe ME, Nilsson A, Verger R. Pancreatic lipase-related protein 2 but not classical pancreatic lipase hydrolyzes galactolipids. Biochim Biophys Acta 1996,1302,236–40.

[87] Yang Y, Sanchez D, Figarella C, Lowe ME. Discoordinate expression of pancreatic lipase and two related proteins in the human fetal pancreas. Pediatr Res 2000,47,184–8.

[88] De Caro J, Sias B, Grandval P, Ferrato F, Halimi H, Carrière F, et al. Characterization of pancreatic lipase-related protein 2 isolated from human pancreatic juice. Biochim Biophys Acta 2004,1701,89–99.

[89] Sias B, Ferrato F, Grandval P, Lafont D, Boullanger P, De Caro A, et al. Human pancreatic lipase-related protein 2 is a galactolipase. Biochemistry 2004,43,10138–48.

[90] Lowe ME, Kaplan MH, Jackson-Grusby L, D'Agostino D, Grusby MJ. Decreased neonatal dietary fat absorption and T cell cytotoxicity in pancreatic lipase-related protein 2-deficient mice. J Biol Chem 1998,273,31215–21.

[91] Roussel A, Yang Y, Ferrato F, Verger R, Cambillau C, Lowe M. Structure and activity of rat pancreatic lipase-related protein 2. J Biol Chem 1998,273,32121–8.

[92] Frossard JL, Steer ML, Pastor CM. Acute pancreatitis. Lancet 2008,371,143–52.

[93] Van Dyke JA, Stanley RJ, Berland LL. Pancreatic imaging. Ann Intern Med 1985,102,212–7.

[94] Moossa AR. Current concepts. Diagnostic tests and procedures in acute pancreatitis. N Engl J Med 1984,311,639–43.

[95] Eckfeldt JH, Levitt MD. Diagnostic enzymes for pancreatic disease. Clin Lab Med 1989,9,731–43.

[96] Ventrucci M, Pezzilli R, Naldoni P, Platè L, Baldoni F, Gullo L, et al. Serum pancreatic enzyme behavior during the course of acute pancreatitis. Pancreas 1987,2,506–9.

[97] Winter JM, Cameron JL, Campbell KA, Arnold MA, Chang DC, Coleman J, et al. 1423 pancreaticoduodenectomies for pancreatic cancer. A single-institution experience. J Gastrointest Surg 2006,10,1199–210.

[98] Winter JM, Cameron JL, Lillemoe KD, Campbell KA, Chang D, Riall TS, et al. Periampullary and pancreatic incidentaloma: a single institution's experience with an increasingly common diagnosis. Ann Surg 2006,243,673–80.

[99] Ghaneh P, Costello E, Neoptolemos JP. Biology and management of pancreatic cancer. Gut 2007,56,1134–52.

[100] Locker GY, Hamilton S, Harris J, Jessup JM, Kemeny N, Macdonald JS, et al. ASCO 2006 update of recommendations for the use of tumor markers in gastrointestinal cancer. J Clin Oncol 2006,24,5313–27.

[101] Bluemke DA, Cameron JL, Hruban RH, Pitt HA, Siegelman SS, Soyer P, et al. Potentially resectable pancreatic adenocarcinoma: spiral CT assessment with surgical and pathologic correlation. Radiology 1995,197,381–5.

[102] Hafkenscheid JC, Hessels M, Jansen JB, Lamers CB. Serum trypsin, alpha-amylase and lipase during bombesin stimulation in normal subjects and patients with pancreatic insufficiency. Clin Chim Acta 1984,136,235–40.

[103] Bobadilla JL, Farrell MH, Farrell PM. Applying CFTR molecular genetics to facilitate the diagnosis of cystic fibrosis through screening. Adv Pediatr 2002,49,131–90.

[104] Bobadilla JL, Macek M Jr, Fine JP, Farrell PM. Cystic fibrosis: a worldwide analysis of CFTR mutations – correlation with incidence data and application to screening. Hum Mutat 2002,19,575–606.

[105] Rowe SM, Miller S, Sorscher EJ. Cystic fibrosis. N Engl J Med 2005,352,1992–2001.

[106] Adriaenssens K, Van Riel L. Serum pancreatic lipase as a screening test for cystic fibrosis. Arch Dis Child 1982,57,553–5.

[107] Eledrisi MS, Alshanti MS, Shah MF, Brolosy B, Jaha N. Overview of the diagnosis and management of diabetic ketoacidosis. Am J Med Sci 2006,331,243–51.

[108] Shenoy SD, Cody D, Rickett AB, Swift PG. Acute pancreatitis and its association with diabetes mellitus in children. J Pediatr Endocrinol Metab 2004,17,1667–70.

[109] Slyper AH, Wyatt DT, Brown CW. Clinical and/or biochemical pancreatitis in diabetic ketoacidosis. J Pediatr Endocrinol 1994,7,261–4.

[110] Quiros JA, Marcin JP, Kuppermann N, Nasrollahzadeh F, Rewers A, DiCarlo J, et al. Elevated serum amylase and lipase in pediatric diabetic ketoacidosis. Pediatr Crit Care Med 2008,9,418–22.

[111] Foster DW, McGarry JD. The metabolic derangements and treatment of diabetic ketoacidosis. N Engl J Med 1983,309,159–69.

[112] Junge W, Malyusz M, Ehrens HJ. The role of the kidney in the elimination of pancreatic lipase and amylase from blood. J Clin Chem Clin Biochem 1985,23,387–92.

[113] Frank B, Gottlieb K. Amylase normal, lipase elevated: is it pancreatitis? A case series and review of the literature. Am J Gastroenterol 1999,94,463–9.

[114] Yadav D, Nair S, Norkus EP, Pitchumoni CS. Nonspecific hyperamylasemia and hyperlipasemia in diabetic ketoacidosis: incidence and correlation with biochemical abnormalities. Am J Gastroenterol 2000,95,3123–8.

[115] Warshaw AL, Feller ER, Lee KH. On the cause of raised serum-amylase in diabetic ketoacidosis. Lancet 1977,1,929–31.

[116] Tietz NW. Support of the diagnosis of pancreatitis by enzyme tests – old problems, new techniques. Clin Chim Acta 1997,257,85–98.

[117] Jiang CF, Ng KW, Tan SW, Wu CS, Chen HC, Liang CT, et al. Serum level of amylase and lipase in various stages of chronic renal insufficiency. Zhonghua Yi Xue Za Zhi (Taipei) 2002,65,49–54.

[118] Seno T, Harada H, Ochi K, Tanaka J, Matsumoto S, Choudhury R, et al. Serum levels of six pancreatic enzymes as related to the degree of renal dysfunction. Am J Gastroenterol 1995,90,2002–5.

[119] Masoero G, Bruno M, Gallo L, Colaferro S, Cosseddu D, Vacha GM. Increased serum pancreatic enzymes in uremia: relation with treatment modality and pancreatic involvement. Pancreas 1996,13,350–5.

[120] Magnusson M, Magnusson KE, Sundqvist T, Denneberg T. Impaired intestinal barrier function measured by differently sized polyethylene glycols in patients with chronic renal failure. Gut 1991,32,754–9.

[121] Arcidiacono R, Gambitta P, Rossi A, Grosso C, Bini M, Zanasi G. The use of a long-acting somatostatin analogue (octreotide) for prophylaxis of acute pancreatitis after endoscopic sphincterotomy. Endoscopy 1994,26,715–8.

[122] Frank CD, Adler DG. Post-ERCP pancreatitis and its prevention. Nat Clin Pract Gastroenterol Hepatol 2006,3,680–8.

[123] Lancashire RJ, Cheng K, Langman MJ. Discrepancies between population-based data and adverse reaction reports in assessing drugs as causes of acute pancreatitis. Aliment Pharmacol Ther 2003,17,887–93.

[124] Thomson SR, Hendry WS, McFarlane GA, Davidson AI. Epidemiology and outcome of acute pancreatitis. Br J Surg 1987,74,398–401.

[125] Anderson V, Carneiro M, Bulterys M, Douglas G, Polliotti B, Slikker W Jr. Perinatal infections: HIV and co-infections in the placenta and therapeutic interventions – a workshop report. Placenta 2001,22 (Suppl A),S34–7.

[126] Balani AR, Grendell JH. Drug-induced pancreatitis: incidence, management and prevention. Drug Saf 2008,31,823–37.

[127] Eland IA, van Puijenbroek EP, Sturkenboom MJ, Wilson JH, Stricker BH. Drug-associated acute pancreatitis: twenty-one years of spontaneous reporting in The Netherlands. Am J Gastroenterol 1999,94,2417–22.

[128] Nitsche CJ, Jamieson N, Lerch MM, Mayerle JV. Drug induced pancreatitis. Best Pract Res Clin Gastroenterol 2010,24,143–55.

[129] Young DS, Pestaner LC, Gibberman V. Effects of drugs on clinical laboratory tests. Clin Chem 1975,21,1D–432D.

[130] Wang CS, Bass HB, Downs D, Whitmer RK. Modified heparin-sepharose procedure for determination of plasma lipolytic activities of normolipidemic and hyperlipidemic subjects after injection of heparin. Clin Chem 1984,27,663–8.

[131] Tietz NW, Astles JR, Shuey DF. Lipase activity measured in serum by a continuous-monitoring pH-stat technique – an update. Clin Chem 1989,35,1688–93.

[132] Scow RO. Effect of sodium taurodeoxycholate, CaCl2 and albumin on the action of pancreatic lipase on droplets of trioleoylglycerol and the release of lipolytic products into aqueous media. Biochimie 1988,70,1251–61.

[133] Sternby B, Akerstrom B. Immunoreactive pancreatic colipase, lipase and phospholipase A2 in human plasma and urine from healthy individuals. Biochim Biophys Acta 1984,789,164–9.

[134] Kropp J, Knapp FF Jr, Weyenberg A, McPherson DW, Ambrose KR, Callahan AP, et al. Evaluation of pancreatic lipase activity by simple urine analysis after oral administration of a new iodine-131-labeled triglyceride. Eur J Nucl Med 1994,21,1227–30.

[135] Fredrikzon B, Olivecrona T. Decrease of lipase and esterase activities in intestinal contents of newborn infants during test meals. Pediatr Res 1978,12,631–4.

[136] Mohiuddin J, Katrak A, Junglee D, Green MF, Dandona P. Serum pancreatic enzymes in the elderly. Ann Clin Biochem 1984,21,102–4.

[137] Lessinger JM, Arzoglou P, Ramos P, Visvikis A, Parashou S, Calam D, et al. Preparation and characterization of reference materials for human pancreatic lipase: BCR 693 (from human pancreatic juice) and BCR 694 (recombinant). Clin Chem Lab Med 2003,41,169–76.

[138] Lessinger JM, Dourson JL, Ferard G. Importance of standardization of lipase assays by using appropriate calibrators. Clin Chem 1996,42,1979–83.

[139] Lessinger JM, Férard G, Mignot V, Calam DH, Das RG, Dourson JL. Catalytic properties and stability of lipase purified from human pancreatic juice. Clin Chim Acta 1996,251,119–29.

[140] Tietz NW, Repique EV. Proposed standard method for measuring lipase activity in serum by a continuous sampling technique. Clin Chem 1973,19,1268–75.

[141] Ballot C, Favre-Bonvin G, Wallach JM. Lipase assay in duodenal juice using a conductimetric method. Clin Chim Acta 1984,143,109–14.

[142] Rood FW, Coleman PF. Solid-phase enzyme immunoassay for the quantitation of pancreatic lipase in human serum. Clin Chem 1986,32,A1129 (abstract).

[143] Shihabi ZK, Bishop C. Simplified turbidimetric assay for lipase activity. Clin Chem 1971,17,1150–3.

[144] Kitamura T, Kurooka S, Tanaka S, Ehara M. Serum lipase response to cerulein secretin stimulation test in patients with pancreas-associated diseases as measured by sensitive colorimetric assay (a BALB-DTNB method). Dig Dis Sci 1984,29,600–5.

[145] Hoffmann GE, Weiss L. Specific serum pancreatic lipase determination, with use of purified colipase. Clin Chem 1980,26,1732–3.

[146] Salvayre R, Nègre A, Radom J, Douste-Blazy L. Fluorometric assay for pancreatic lipase. Clin Chem 1986,32,1532–6.

[147] Bartl K, Brandhuber M, Ziegenhorn J. Improved automated kinetic determination of uric acid in serum by use of uricase/catalase/aldehyde dehydrogenase. Clin Chem 1979,25,619–21.

[148] Fossati P, Ponti M, Paris P, Berti G, Tarenghi G. Kinetic colorimetric assay of lipase in serum. Clin Chem 1992,38,211–5.

[149] Melzi d ' Eril GV, Bosoni T, Moratti R, Ventrucci M, Fumagalli A, Tarenghi G. Clinical validity of a continuous colorimetric method for serum lipase. Eur J Clin Chem Clin Biochem 1992,30,439–44.

[150] Panteghini M, Pagani F, Bonora R. Clinical and analytical evaluation of a continuous enzymatic method for measuring pancreatic lipase activity. Clin Chem 1993,39,304–8.

[151] Grenner G, Deutsch G, Schmidtberger R, Dati F. A highly sensitive enzyme immunoassay for the determination of pancreatic lipase. J Clin Chem Clin Biochem 1982,20,515–9.

[152] Lott JA, Patel ST, Sawhney AK, Kazmierczak SC, Love JE Jr. Assays of serum lipase: analytical and clinical considerations. Clin Chem 1986,32,1290–302.

[153] Møller-Petersen J, Klaerke M, Dati F, Toth T. Immunochemical qualitative latex agglutination test for pancreatic lipase in serum evaluated for use in diagnosis of acute pancreatitis. Clin Chem 1985,31,1207–10.

[154] Roberts IM, Mercer D. Radioimmunoassay for human pancreatic lipase in acute pancreatitis. Dig Dis Sci 1987,32,388–92.

10 Natriuretic peptides

Robert H. Christenson and Hassan M.E. Azzazy

10.1 Case studies

10.1.1 Patient A (Is B-type natriuretic peptide a better test for rule in or rule out of congestive heart failure?)

A 47-year-old female presents to the emergency department (ED) with dyspnea as her main complaint. The patient smokes one pack of cigarettes per day and has moderate ethanol use of two mixed drinks per day. The patient has no history of cardiovascular disease or heart failure (HF), but had asthma attacks as a teenager. The patient is examined by a first-year emergency medicine resident who is aware that B-type natriuretic peptide (BNP) can be a useful marker for the workup of congestive heart failure (CHF). After performing a physical examination and taking the patient's history, the resident thinks there is a 50% chance that CHF is the cause of the patient's dyspnea. He requests a BNP test and wonders about his interpretation if the test is positive or negative. He is curious whether BNP is a better rule in or rule out test.

10.1.1.1 Discussion

Use of Bayes' theorem, that is, post-test odds = pre-test odds × likelihood ratio, is standard for interpretation of clinical diagnostic tests. A robust way to retrieve likelihood ratios is from meta-analysis studies that may be available for the test. Relevant to this case and the BNP question, a recent BNP meta-analysis was conducted by the UK National Institute for Clinical Excellence (Craig J, et al. Health technology assessment report 6. Glasgow, UK, NHS Quality Improvement Scotland, 2005, www.nhshealthquality.org.). This meta-analysis reported a positive likelihood ratio of 3.02 and a negative likelihood ratio of 0.15. Rather than using the pre-test odds, for general interpretation the odds are expressed in terms of pre-test probability because this is viewed as more straightforward and convenient. Use of a tool termed Fagan's nomogram allows direct conversion of pre-test probability to post-test probability when likelihood ratios are known (this tool can be found at <http://www.cebm.net/index.aspx?o=1161>).

To answer this question, the physician goes to the above website with the Fagan's diagram tool. He starts by moving the arrow at the pre-test probability (left-most) line to 50%; he then places the arrow for the positive likelihood ratio (central line) at 3.02. He then reads the post-test probability off the right-most line as 77%. Thus, if the patient has a positive BNP result (>100 ng/L), the physician's certainty that this patient has CHF is raised from the pre-test value of 50% to 77%. By contrast, if the BNP test is negative (<100 ng/L), the pre-test probability is also 50% (left-most line); the arrow for the negative likelihood ratio is positioned at 0.15 (center line). Here, the post-test probability is 10%, which is a five-fold improvement in certainty from the pre-test probability of 50%. Therefore, a positive test still leaves 24% uncertainty for ruling in CHF, whereas a negative test leaves only a 10% uncertainty for ruling out CHF. For this reason, BNP may be considered a better rule out test than it is a rule in test for CHF.

10.1.2 Patient B (diagnosis of CHF)

A 77-year-old Caucasian female in acute distress presents to the ED with dyspnea as her main complaint. She is unable to breathe easily unless sitting up straight (ortho-pnea) and reports recent fatigue and effort intolerance. On physical examination she has afebrile and has no jugular venous distension, peripheral edema, hepatospleno-megaly, or apparent ascites. Her medical history is significant for a myocardial infarc-tion 5 years previously. The patient's estimated glomerular filtration rate (eGFR) was >60 mL/min/1.73 m^2 based on her creatinine concentration, age, and race. The atten-ding physician suspected that the patient had acute decompensated heart failure (ADHF) and requested an N-terminal pro-B-type natriuretic peptide (NT-proBNP) level. The NT-proBNP result was 2,500 ng/L.

10.1.2.1 Discussion
The normal reference interval for NT-proBNP is 125 ng/L for individuals under 75 years old and 450 ng/L for individuals aged 75 years old and older. However, use of normal reference interval as the decision limit is inappropriate in acutely ill patients presen-ting with signs and symptoms of HF. The International Collaborative of NT-proBNP Study [1] demonstrated stratification of NT-proBNP using cut-off points of 450, 900, and 1,800 ng/L for age groups of <50, 50–75, and >75 years, respectively. Use of these age-dependent cut-off points reduces false-negative findings in younger patients, reduces false-positive findings in older patients, and improves the overall positive predictive value (PPV) of the marker without a change in overall sensitivity or specifi-city [2]. It is noteworthy that a single decision point of 100 ng/L is used for BNP regard-less of age [3]. The patient's NT-proBNP result is consistent with a diagnosis of ADHF.

10.1.3 Patient C

A 65-year-old male presents with acute distress to the ED with dyspnea as his main complaint. He is unable to breathe easily unless sitting up straight (orthopnea) and reports recent fatigue and effort intolerance. On physical examination he is afebrile and has no jugular venous distension, peripheral edema, hepatosplenomegaly, or apparent ascites. The patient is a type 2 diabetic and his medical history is significant for chronic kidney disease (CKD) with an eGFR that is typically 35 mL/min/1.73 m^2, and he had a myocardial infarction 4 years previously. The attending physician suspected that the patient has ADHF based on his history, signs, and symptoms and requested an NT-proBNP level. The NT-proBNP result was 3,500 ng/L.

10.1.3.1 Discussion

NT-proBNP levels are typically higher in patients with CKD. Although modestly stronger, inverse correlations exist between renal function and NT-proBNP compared with BNP; both biomarkers are dependent on renal clearance to a similar degree [4]. Elevated NT-proBNP levels do not simply reflect the reduced renal clearance of the peptide, but are believed to also reflect the increased prevalence of heart disease in CKD patients. Correlations between BNP and NT-proBNP remain strong across the continuum of CKD. When evaluating the patient with acute dyspnea and CKD, both BNP and NT-proBNP are affected similarly, with higher decision limits necessary compared with patients with preserved renal function. An age-independent NT-proBNP cut-off point of 900 ng/L has a similar value as that reported for a BNP value of 100 ng/L [2]. In renal insufficiency patients, higher cut-off points of 1,200 ng/L for NT-proBNP and 200 ng/L for BNP were documented in an all-comers dyspnea population [4]. Importantly, when using NT-proBNP to evaluate a patient with dyspnea and impaired renal function, the recommended cut-off points of 450, 900, and 1,800 ng/L for those aged <50, 50–75, and >75 years, respectively, do not require further adjustment for renal function [5]; BNP and NT-proBNP are useful for the diagnostic evaluation of patients with CKD [5].

10.1.4 Patient D (natriuretic peptides and obesity)

A 79-year-old female presents in acute distress to the ED with dyspnea as her main complaint. The patient is obese, she is 1.5 m tall (4 feet and 11 inches) and weighs 91 kg (200 lbs) and therefore has a body mass index (BMI) of 40.4 kg/m^2. She reports recent fatigue and is febrile with a temperature of 38.4°C (101°F). The patient has an eGFR of >60 mL/min/1.73 m^2 and has no history of coronary artery disease. The attending physician suspects that the patient's symptoms could be due to ADHF and requests a BNP level, which is 90 ng/L (positive cut-off point: 100 ng/L). The attending physician knows that BNP is a powerful rule out test and wonders if this patient could still have HF.

10.1.4.1 Discussion

The World Health Organization defines obesity according to BMI, with a value ≤25 kg/m^2 being normal, BMI between 25 and 30 kg/m^2 overweight, and BMI ≥30 kg/m^2 obese. Surprisingly, for reasons that remain unclear, obese patients with HF have a better prognosis than patients whose weight is normal, giving rise to the so-called obesity paradox [6]. Also, levels of both BNP and NT-proBNP are relatively lower in patients with a higher BMI than in those with a lower BMI [7]. Based on available data, the lower values associated with obesity are more likely to be related to reduced synthesis or secretion of the peptides rather than increased clearance (which may play only a minor role in this context) [8]. In a recent study examining BNP and BMI, cut-off points for 90% diagnostic sensitivity for a HF diagnosis were 170 ng/L for lean patients, 110 ng/L for overweight/obese patients, and a much lower 54 ng/L in severely/morbidly obese patients [9]. Therefore, the BNP value of 90 ng/L for this obese patient is well above the 90% sensitivity cut-off point of 54 ng/mL, and although this is below the similar cut-off points for normal or overweight individuals, the attending physician must take this into consideration when determining the diagnosis for this obese patient.

It is interesting that age-adjusted NT-proBNP cut-off points to "rule in" HF and age-independent cut-off points to "rule out" HF in patients with acute dyspnea are equally useful for obese and lean patients, and no adjustment of NT-proBNP thresholds for BMI is recommended [8].

10.1.5 Patient E (NT-proBNP for long-term risk assessment of elderly non-HF individuals)

An ambulatory 73-year-old male, who is in good general health and lives in the community, has a wellness appointment with his general practitioner. This patient has no signs and symptoms of HF, but his practitioner knows that subclinical cardiovascular disease is common in elderly people and is associated with an increased risk of cardiovascular events, including HF [10] and occult HF. The general practitioner recently attended a continuing education activity that discussed BNP and NT-proBNP. The general practitioner recalls that the speaker noted that these are indicators of hemodynamic stress and wonders if they can be used in his practice to screen for individuals who are likely to develop HF.

10.1.5.1 Discussion

Although it is not a standard of care, NT-proBNP has been investigated for assessment of risk for new-onset HF and death from a cardiovascular cause of HF in free-living elderly patients [11]. When approximately 3,000 elderly patients from the Cardiovascular Health Study were divided into quintiles based on NT-proBNP values, patients in the highest quintile (>268 ng/L) were at three-fold risk of developing HF and a three-fold risk of cardiovascular death compared with individuals in the lowest quintile (<47.5 ng/L). Elevated risk began to occur at an NT-proBNP level of 190 ng/L. Among participants with initially low NT-proBNP (<190 ng/L), those who developed a >25% increase in the following 2–3 years later were at two-fold greater adjusted risk of HF and cardiovascular death compared with those with sustained levels <190 ng/L. Among participants with initially high NT-proBNP results, those who developed a >25% NT-proBNP increase were also at two-fold higher risk of HF and cardiovascular death, whereas those who developed a <25% decrease to <190 ng/L were at approximately one-half the risk of HF and cardiovascular death compared with those with unchanged high values [11].

The general practitioner could not justify the cost of the NT-proBNP measurement for this patient on this visit. In the future, however, this biomarker may be an important component of evaluating elderly patients for HF risk and cardiovascular events.

10.1.6 Patient F (prognosis in chronic HF)

An 83-year-old Caucasian female was diagnosed with HF 5 years previously. Although this patient's chronic HF has apparently responded well to her physician's management strategy, her physician questions her long-term prognosis. The patient has had NT-proBNP performed at each of her outpatient visits for the past 2 years, which have remained stable at approximately 1,500 ng/L.

10.1.6.1 Discussion

NT-proBNP and BNP levels are among the strongest independent predictors of adverse events in patients with chronic HF, and their measurement is useful for prognostication across the continuum of HF disease [12, 13]. In patients with chronic HF, repeated determinations of NT-proBNP or BNP appear to provide additional prognostic value for relevant outcomes, including death or HF hospitalization. Although precise target values for NT-proBNP values are not currently defined, morbidity and mortality in chronic HF appear to increase markedly with an NT-proBNP concentration >1,000 ng/L [12] or BNP value >334 ng/L [4]. Confounding factors including renal insufficiency or obesity must be considered when prognostically evaluating patients using NT-proBNP or BNP measurements, but the value of NT-proBNP or BNP is also relevant in these patients. Serial measurements of NT-proBNP or BNP may be valuable for prognostication in chronic HF and a measurement at each patient visit or after of changes in clinical stability may be a useful prognostic aid.

10.2 Biochemistry and physiology of cardiac natriuretic peptides

The cardiac natriuretic peptides include atrial natriuretic peptide (ANP) and B-type natriuretic peptide (BNP), and their related metabolic peptides. BNP and NT-proBNP have been found to be more predictably associated with disease than ANP, thus development of ANP clinical assays has not been pursued for routine clinical use. BNP is synthesized inside myocytes as a 108 amino acid (aa) precursor molecule. Upon stimulus for release, this precursor is enzymatically cleaved into the active 32 aa BNP hormone (with a ring structure) and a metabolically inactive N-terminal protein having 76 aa (NT-proBNP). Key characteristics of BNP and NT-proBNP are shown in Table 10.1 [14].

Table 10.1: Comparison between BNP and NT-proBNP.

	BNP	NT-proBNP
Structure	32 aa	76 aa
Physiological activity	Yes	No
Synthesis	Cleavage from proBNP (108 aa)	Cleavage from proBNP (108 aa)
Half-life	13–20 min	25–70 min
Clearance	Neutral peptidases	Renal clearance
	Clearance receptors	
	Renal clearance	

Table 10.2: Values for BNP and NT-proBNP.

	BNP	NT-proBNP
Normal levels	Men: 20 ng/L	<75 years: 125 ng/L
	Women: 25 ng/L	>75 years: 450 ng/L
Cut-off for HF diagnosis	100 ng/L	<50 years: 450 ng/L
		50–75 years: 900 ng/L
		>75 years: 1,800 ng/L
Rule out values	50 ng/L	<300 ng/L

The natriuretic peptides are released in response to increased hemodynamic stress and wall tension. Their main physiological actions include regulation of blood pressure, fluid balance, and sodium excretion. Additionally, ANP/BNP are involved in inhibition of the sympathetic nervous system, counterbalance of the renin-angiotensin-aldosterone system, and inhibition of the endothelins, vasopressin, and cytokines; the natriuretic peptides also appear to play a role in the inhibition of ventricular and vascular hyper-trophy and remodeling. The natriuretic peptides have beneficial effects on endothelial dysfunction subsequent to atherosclerosis such as blunting of shear stress, regulation of coagulation and fibrinolysis, and inhibition of platelet coagulation [14].

The plasma concentration of BNP is controlled by the rate of synthesis and release, and balanced by clearance mechanisms which include active receptors (found in vascular endothelium, smooth muscle, heart, and kidney), enzymatic degradation in circulation, and receptor-mediated endocytosis. By contrast, NT-proBNP is cleared solely by glomerular filtration. Plasma values for BNP and NT-proBNP are listed in Table 10.2.

10.3 Chemical pathology of natriuretic peptides

Natriuretic peptides are increased in all disorders with salt and fluid overload and those with increased atrial or ventricular wall tension such as acute or chronic systolic or diastolic left and right HF. They may also be elevated in valvular heart disease, atrial fibrillation, pulmonary embolism, inflammatory cardiac disease, acute or chronic renal failure, anemia, trauma, sepsis, and stroke. For both markers, only significant changes from baseline values are related to clinical outcomes [15].

10.3.1 BNP in diagnosis of symptomatic HF patients

The value of BNP and NT-proBNP in aiding the diagnosis of HF in patients presenting with dyspnea is well established. In patients with acute dyspnea without severe renal failure, a BNP value of <100 ng/L or NT-proBNP <300 ng/L were found to be optimal to rule out HF. BNP values >500 ng/L or NT-proBNP >450 ng/L (<50 years), >900 ng/L (50–75 years), or 1,800 ng/L (>75 years) are suggestive of HF (pending confirmation by echocardiography) [15, 16].

The "Breathing Not Properly" Study was the first large-scale multicenter, multinational, prospective study to evaluate the use of BNP for diagnosis of HF in patients admitted to ED with acute dyspnea [17]. The ED physicians, blinded to BNP results, were asked to evaluate the probability of patients having HF. Two cardiologists, also blinded to BNP levels, were additionally asked to review all clinical data and standardize scores to make a diagnosis. Compared with patient history, physical measurements, and other biochemical results, BNP levels were the most accurate predictor of HF diagnosis. BNP levels were elevated in patients with subsequent HF (675 ng/L) compared with those with non-cardiac dyspnea (110 ng/L). A BNP cut-off of 100 ng/L differentiated HF from other causes of dyspnea (90% sensitivity and 76% specificity). A BNP cut-off of 50 ng/L had a negative predictive value (NPV) of 96%.

The clinical utility of BNP and NT-proBNP in the diagnosis of HF was compared in 205 patients presenting to the ED with acute dyspnea [18]. Diagnostic classification of the two assays correlated well (r = 0.902, p < 0.0001). The best sensitivities and specificities were achieved at a BNP value of 208 ng/L (specificity 70%, sensitivity 94%, PPV 61%, NPV 96%, accuracy 78%) and an NT-proBNP value of 30 pmol/L (specificity 87%, sensitivity 80%, PPV 76%, NPV 89%, accuracy 85%). The areas under the receiver operator characteristic (ROC) curve were identical for BNP and NT-proBNP in the population. Several age-based NT-proBNP optimal cut-off points to rule in acute CHF diagnosis were proposed (Table 10.2). The rule out acute CHF cut-off of 300 ng/L had 99% sensitivity, 62% specificity, 55% PPV, 99% NPV, and 83% accuracy.

10.3.2 Prognosis and risk stratification of HF patients

It is well established that high values of the natriuretic peptides are associated with an increased probability of adverse events. The Valsartan in Heart Failure Trial (Val-HeFT) evaluated outcomes in patients with moderate to severe HF [19]. BNP was measured in all patients at enrolment, and then at 4, 12, and 24 months. In comparison with patients with BNP levels below the median, patients with BNP levels above the median had a relative risk of 2.1 for mortality and 2.2 for first morbid events.

Additionally, patients in the lowest quartile of BNP (<41 ng/L) had the lowest all-cause mortality, whereas patients in the highest BNP quartile (>238 ng/L) had significantly higher mortality of 32% at 30 months [19].

10.3.3 BNP and NT-proBNP as prognostic risk markers in acute coronary syndrome patients

Acute coronary syndrome (ACS) represents a spectrum of diseases that covers unstable angina through frank myocardial infarction (MI). In ACS, ischemia results in local myocardial dysfunction and increased hemodynamic stress and wall tension. In frank MI patients, necrosis also causes increased hemodynamic stress, reduced cardiac performance, and symptoms of ventricular dysfunction. Patients having increased hemodynamic stress during their acute myocardial event will have a greater probability of ventricular remodeling that can cause progressive left ventricular dilatation and dysfunction with development of HF. Further, these patients will also have a greater number of other adverse outcomes. Hemodynamic stress and ventricular wall tension increase BNP and NT-proBNP, thus these biomarkers can be used for risk assessment and prognosis of ACS patients [20].

10.3.4 Recommendations for use of biochemical markers in heart failure

Recently, the American College of Cardiology Foundation and American Heart Association have issued recommendations for use of biochemical markers in heart failure (Table 3) [21]. In addition to natriuretic peptides (NP), other biochemical markers of myocardial remodeling, inflammation, oxidative stress, and neurohormonal imbalance have been investigated for their prognostic value in HF. Soluble ST2 and galectin-3 (markers of myocardial fibrosis) were found to be predictive of hospitalization and death and have demonstrated an additive prognostic value to natriuretic peptides (NP) in HF patients [21].

10.4 Analytical measurements of BNP and NT-proBNP

10.4.1 Desired specimens

For BNP, EDTA anticoagulated whole blood or plasma, ideally collected in plastic iced tubes, are acceptable specimens. The preferred specimen for NT-proBNP is serum or heparin plasma collected in plastic or glass tubes. BNP should be assayed within 4 h after blood collection to avoid *in vitro* degradation. Stability of NT-proBNP is higher than that of BNP.

Table 10.3: Recommendations for biomarkers in HF.

Biomarker, Application	Setting	COR	LOE
Natriuretic peptides			
Diagnosis or exclusion of HF	Ambulatory, acute	I	A
Prognosis of HF	Ambulatory, acute	I	A
Achieve GDMT	Ambulatory	IIa	B
Guidance for acutely decompensated HF therapy	Acute	IIb	C
Biomarkers of myocardial injury			
Additive risk stratification	Acute, ambulatory	I	A
Biomarkers of myocardial fibrosis (sST2 and Galectin-3)			
Additive risk stratification	Ambulatory	IIb	B
	Acute	IIb	A

COR indicates Class of Recommendation; GDMT, guideline-directed medical therapy; HF, heart failure; LOE, Level of Evidence. Level A: multiple populations evaluated; Level B: limited populations evaluated; Level C: very limited populations evaluated, only consensus opinion of experts, case studies, or standard of care. Class I: Procedure/treatment should be administered. Class IIa: additional studies with focused objectives are needed. Class IIb: additional studies with broad objectives are needed; additional registry data would be helpful. Adapted with permission from reference [21].

10.4.2 Considerations for measurements of BNP and NT-proBNP

10.4.2.1 Age and gender
Plasma levels of BNP and NT-proBNP increase with age and could be due to a decline in renal function and reduced clearance of these peptides. Both peptides are higher in women than in men. Although age and gender differences are greater for NT-proBNP than BNP, it has been suggested that age- and gender-specific reference intervals should be considered in the routine interpretation of results of both assays (Table 10.2).

10.4.2.2 Renal insufficiency
BNP and NT-proBNP are released in an equimolar manner; however, their levels in circulation may differ substantially due to substantial differences in half-lives and clearance mechanisms. Both peptides are excreted by the kidneys and the assay cut-off value distinguishing normal and pathological conditions is sensitive to renal insufficiency. Higher cut-off points must be considered when evaluating patients with renal insufficiency.

10.4.2.3 Obesity
Obesity is a major risk factor for hypertension and other disorders associated with HF. Sodium retention and excessive sympathetic tone are frequently related to obesity,

and BNP is an important regulator of sodium homeostasis and neurohormonal activation. For every kilogram of excess weight, there are several miles of vasculature that are abundant in BNP receptors. Because NT-proBNP is not cleared by a receptor-mediated mechanism, obese individuals may have reduced BNP levels compared with levels of NT-proBNP.

The relationship between plasma levels of BNP and BMI was investigated in 3,389 non-HF individuals (1,803 women) from the Framingham Study [22]. Compared with lean individuals (BMI <25 kg/m^2), obese individuals (BMI ≥30 kg/m^2) had higher odds of having low BNP plasma levels [multivariate-adjusted odds ratio: men 2.51 (95% confidence interval: 1.71–3.68); women 1.84 (95% confidence interval: 1.32–2.58)]. In addition to implications for the use of BNP levels in patient care, such low circulating levels of BNP may contribute to the susceptibility of obese individuals to hypertension and related disorders such as left ventricular hypertrophy.

Christenson et al. [23] measured both peptides in 685 patients (280 obese, 193 overweight, and 212 normal BMI) with possible decompensated HF in a free-living community population. Obese patients had lower BNP and NT-proBNP compared with overweight or normal-weight individuals (p < 0.001) and decreased mortality compared with normal-weight individuals (p < 0.001). NT-proBNP outperformed BNP for predicting all-cause mortality in normal-weight individuals. Using multivariate regression, both biomarkers remained significant predictors of decompensated HF diagnosis in each BMI subgroup.

10.4.2.4 Diabetes

In a preliminary study, plasma NT-proBNP levels were measured in patients from primary care centers who had no overt heart disease: 253 patients with type 2 diabetes and 230 matched controls. NT-proBNP levels were higher in type 2 diabetics without overt heart disease (361 pmol/L) compared with controls (303 pmol/L) (p < 0.001). Measurement of NT-proBNP, if paired with cystatin C to estimate glomerular filtration rate, might be used to identify diabetic patients at risk for ventricular dysfunction and who could benefit from an echocardiographic examination [24].

10.4.3 Methods for measurements

Several BNP and NT-proBNP immunoassays are commercially available and have been approved by the US Food and Drug Administration (FDA) to aid diagnosis of HF. Clinical studies have documented the clinical utility of such assays for the differential diagnosis of dyspnea, risk stratification of patients with HF and ACS, and detection of left ventricular systolic and/or diastolic dysfunction [25, 26]. The Shinogi BNP test, which was approved in 2003 and now available in automated platforms, was used in most early clinical studies. The Triage BNP employs one fluorescently labeled polyclonal antibody and an immobilized monoclonal antibody that recognizes the ring structure of BNP. This test, with a turnaround time of 15 min, has been used by most recent clinical studies investigating CHF and ACS patients. It should be noted that current natriuretic peptide assays are not standardized and thus results of a given patient obtained by different assays are not comparable. Additionally, different BNP and NT-proBNP assays which employ the same antibodies may generate different results for a given specimen due to matrix effects.

10.5 Questions and answers

1. What are the major physiological functions of natriuretic peptides?
    (a) Natriuresis and diuresis
    (b) Increase in endothelial permeability
    (c) Inhibition of the rennin-angiotensin-aldosterone system
    (d) Inhibition of cardiac and vascular remodeling
    (e) All of the above
2. Plasma levels of BNP and NT-proBNP are influenced by
    (a) Age
    (b) Gender
    (c) Obesity
    (d) Renal insufficiency
    (e) All of the above
3. Which of the following methods are used for measuring natriuretic peptides?
    (a) Spectrophotometry
    (b) High performance liquid chromatography
    (c) Immunoassays
    (d) Gas chromatography
    (e) Capillary electrophoresis
4. Which of the following statements is incorrect with regard to BNP?
    (a) BNP is a hormone which is synthesized in myocytes as a larger inactive molecule proBNP
    (b) BNP is released in response to myocardial wall stress
    (c) A single BNP measurement can be used to help differentiate the cause of acute dyspnea in an emergency setting
    (d) Intermediate elevations in BNP levels can help to rule in acute heart failure syndrome
    (e) The half-life of BNP is blood is approximately 20 min
5. Plasma levels of BNP and NT-proBNP can be used
    (a) To aid diagnosis of acute decompensated heart failure
    (b) As predictors of mortality in ACS patients
    (c) As predictors of mortality in heart failure patients

(d) As predictors of mortality in patients with end stage renal disease

(e) All of the above

6. Which of the following BNP values indicates a high likelihood (PPV = 90%) of heart failure?

(a) 50 ng/L

(b) 100 ng/L

(c) 200 ng/L

(d) 400 ng/L

(e) >500 ng/L

7. Which of the following NT-proBNP cut-off values can be best used to rule out heart failure?

(a) <100 ng/L

(b) <200 ng/L

(c) <300 ng/L

(d) 900 ng/L

(e) 1,800 ng/L

8. Which of the following diseases may cause increased levels of natriuretic peptides?

(a) Acute or chronic heart failure and inflammatory cardiac disease

(b) Acute or chronic renal failure

(c) Atrial fibrillation

(d) Pulmonary embolism

(e) All of the above

References

[1] Januzzi JL, van Kimmenade R, Lainchbury J, Bayes-Genis A, Ordonez-Llanos J, Santalo-Bel M. NT-proBNP testing for diagnosis and short-term prognosis in acute destabilized heart failure: an international pooled analysis of 1256 patients: the International Collaborative of NT-proBNP Study. Eur Heart J 2006,27,330–7.

[2] Januzzi JL, Chen-Tournoux AA, Moe G. Amino-terminal pro-B-type natriuretic peptide testing for the diagnosis or exclusion of heart failure in patients with acute symptoms. Am J Cardiol 2008,101,29–38.

[3] Maisel AS, Krishnaswamy P, Nowak RM, McCord J, Hollander JE, Duc P. Rapid measurement of B-type natriuretic peptide in the emergency diagnosis of heart failure. N Engl J Med 2002,347,161–7.

[4] de Filippi CR, Seliger SL, Maynard S, Christenson RH. Impact of renal disease on natriuretic peptide testing for diagnosing decompensated heart failure and predicting mortality. Clin Chem 2007,53,1511–9.

[5] de Filippi C, van Kimmenade RR, Pinto YM. Amino-terminal pro-B-type natriuretic peptide testing in renal disease. Am J Cardiol 2008,101,82–8.

[6] Horwich TB, Fonarow GC, Hamilton MA, MacLellan WR, Woo MA, Tillisch JH. The relationship between obesity and mortality in patients with heart failure. J Am Coll Cardiol 2001,38,789–95.

[7] Christenson RH, Azzazy HM, Duh S, Maynard S, Seliger SL, Defilippi CR. Impact of increased body mass index on accuracy of B-type natriuretic peptide (BNP) and N-terminal proBNP for diagnosis of decompensated heart failure and prediction of all-cause mortality. Clin Chem 2010,56,633–41.

[8] Bayes-Genis A, de Filippi C, Januzzi JL. Understanding amino-terminal pro-B-type natriuretic peptide in obesity. Am J Cardiol 2008,101,89–94.

[9] Daniels LB, Clopton P, Bhalla V, Krishnaswamy P, Nowak RM, McCord J, et al. How obesity affects the cut-points for B-type natriuretic peptide in the diagnosis of acute heart failure. Results from the Breathing Not Properly Multinational Study. Am Heart J 2006,151,999–1005.

[10] Kuller LH, Arnold AM, Psaty BM, Robbins JA, O'Leary DH, Tracy RP, et al. 10-year follow-up of subclinical cardiovascular disease and risk of coronary heart disease in the Cardiovascular Health Study. Arch Intern Med 2006,166,71–8.

[11] de Filippi CR, Christenson RH, Gottdiener JS, Kop WJ, Seliger SL. Dynamic cardiovascular risk assessment in elderly people. The role of repeated N-terminal pro-B-type natriuretic peptide testing. J Am Coll Cardiol 2010,55,441–50.

[12] Masson S, Latini R. Amino-terminal pro-B-type natriuretic peptides and prognosis in chronic heart failure. Am J Cardiol 2008,101,56–60.

[13] Boerrigter G, Costello-Boerrigter LC, Burnett JC. Natriuretic peptides in the diagnosis and management of chronic heart failure. Heart Fail Clin 2009,5,501–14.

[14] Azzazy HM, Christenson RH. B-type natriuretic peptide: physiologic role and assay characteristics. Heart Fail Rev 2003,8,315–20.

[15] Thygesen K, Mair J, Mueller C, Huber K, Weber M, Plebani M, et al. Study Group on Biomarkers in Cardiology of the ESC Working Group on Acute Cardiac Care. Recommendations for the use of natriuretic peptides in acute cardiac care: a position statement from the Study Group on Biomarkers in Cardiology of the ESC Working Group on Acute Cardiac Care. Eur Heart J 2012,33,2001–6.

[16] Januzzi JL Jr, Camargo CA, Anwaruddin S, Baggish AL, Chen AA, Krauser DG, et al. The N-terminal Pro-BNP investigation of Dyspnea in the Emergency Department (PRIDE) Study. Am J Cardiol 2005,95,948–54.

[17] McCullough PA, Duc P, Omland T, McCord J, Nowak RM, Hollander JE, et al. B-Type natriuretic peptide and renal function in the diagnosis of heart failure: an analysis from the Breathing Not Properly Multinational Study. Am J Kidney Dis 2003,41,571–9.

[18] Lainchbury JG, Campbell E, Frampton CM, Yandle TG, Nicholls MG, Richards AM. Brain natriuretic peptide and N-terminal brain natriuretic peptide in the diagnosis of heart failure in patients with acute shortness of breath. J Am Coll Cardiol 2003,42,728–35.

[19] Anand IS, Fisher LD, Chiang YT, Latini R, Masson S, Maggioni AP, et al. Changes in brain natriuretic peptide and norepinephrine over time and mortality and morbidity in the Valsartan Heart Failure Trial (Val HeFT). Circulation 2003,107,1278–83.

[20] Wiviott SD, de Lemos JA, Morrow DA. Pathophysiology, prognostic significance and clinical utility of B-type natriuretic peptide in acute coronary syndromes. Clin Chim Acta 2004,346, 119–28.

[21] Yancy CW, Jessup M, Bozkurt B, Butler J, Casey DE, Drazner MH, et al. 2013 ACCF/AHA Guideline for the Management of Heart Failure: A Report of the American College of Cardiology Foundation/American Heart Association Task Force on Practice Guidelines. Circulation 2013,128,e240–e327.

[22] Wang TJ, Larson MG, Levy D, Benjamin EJ, Leip EP, Wilson PW, et al. Impact of obesity on plasma natriuretic peptide levels. Circulation 2004,109,594–600.

[23] Christenson RH, Azzazy HM, Duh SH, Maynard S, Seliger SL, Defilippi CR. Impact of increased body mass index on accuracy of B-type natriuretic peptide (BNP) and N-terminal proBNP for diagnosis of decompensated heart failure and prediction of all-cause mortality. Clin Chem 2010,56,633–41.

[24] Magnusson M, Melander O, Israelsson B, Grubb A, Groop L, Jovinge S. Elevated plasma levels of Nt-proBNP in patients with type 2 diabetes without overt cardiovascular disease. Diabetes Care 2004,27,1929–35.

[25] Clerico A, Emdin M. Diagnostic accuracy and prognostic relevance of the measurement of cardiac natriuretic peptides: a review. Clin Chem 2004,50,33–50.

[26] Munagala VK, Burnett JC Jr, Redfield MM. The natriuretic peptides in cardiovascular medicine. Curr Probl Cardiol 2004,29,707–69.

Index

www.ingramcontent.com/pod-product-compliance
Lightning Source LLC
Chambersburg PA
CBHW081104220326
41598CB00038B/7222